# Video Security Systems

# 2nd Edition

KEITH W. BOSE

**BUTTERWORTHS**

Boston • London

**Library of Congress Cataloging in Publication Data**

Bose, Keith W.
   Video security systems.

   Bibliography: p.
   Includes index.
   1. Electronic security systems. 2. Closed circuit television. I. Title.
TH9737.B67   1982        621.388'9        82-1343
ISBN 0-409-95057-2                        AACR2


Published by Butterworth Publishers
10 Towers Office Park
Woburn, MA 01801

10  9  8  7  6  5  4  3  2

*Printed in the United States of America*

# Contents

# List of Illustrations

# *Preface*

This is a revision of the *Video Security Systems*, first published in 1976 by Howard W. Sams & Company. Additional material has been added to help designers, security officers, and others who use the television camera as an important surveillance device. The use of closed-circuit television for surveillance and industrial monitoring has burgeoned in recent years and has now become a billion–dollar–plus yearly industry. The introduction of lower priced high grade lenses, versatile cameras, and video tape recorders has helped this growth. Video surveillance is now an industry unto itself with specialized equipment and personnel in engineering, manufacture, installation, sales, and operation throughout the world.

This book presents the field of video surveillance from the standpoint of the state of the art. It will be found useful for anyone interested in this growing field. The architect or planner will find material of help in procurement and layout. Security officers and law enforcement personnel may benefit from descriptions of installations and operation. The engineer or technician who desires specialized knowledge will find helpful material here. In preparing the material, the author has attempted to present information that is otherwise not available—information developed from recent advances in the field or ordinarily available only to specialists.

The television camera makes it possible for a single pair of eyes to be in many places at once. Cables may link cameras from a distance of many miles to a central point, from which protective services may be dispatched. Little specialized training is necessary to operate a CCTV system once it is installed, and operators soon learn to interpret observations.

Nevertheless, funds have been wasted in the past on video surveillance installations that lack effectiveness and result in excess repair costs. This revised edition includes the theory of video surveillance as well as the technology of Closed-Circuit Television, and will help to gain optimum results from video surveillance systems.

The author and publishers wish to thank the many manufacturers of video equipment who have contributed to this present edition, as well as the past edition. Their contribution has been both direct and indirect, and in areas too numerous to mention.

Keith W. Bose
Kings Park, NY

# 1

# *Closed-Circuit Television— A Tool of Many Uses*

Everyone is familiar with home television, but there is another kind of television that can be used anywhere by anyone for many purposes, and new uses are found every day. Closed-circuit television—both black-and-white and color—can be sent over privately owned cables for distances from a few feet to hundreds of miles. There are no licensing requirements for the typical private installation.

A closed circuit television (CCTV) system operates in the same manner, and in fact is interchangeable electronically with standard broadcast television. In cases where higher standards need not be met, cameras for CCTV can be relatively inexpensive. The advent of solid-state integrated circuitry has resulted in more compact, less expensive cameras and monitors. Newer, faster lenses are now mass produced that can add considerable versatility to television cameras.

1

A leading new growth industry, CCTV is used for remote observation in industry, retail sales, medical care, education, and other diverse fields. Mass production of small TV cameras and specialized lenses has helped this growth by reducing cost and adding both versatility and availability. Along with the camera, lens, and monitor, many accessory devices are needed to provide a complete system. The system shown in Figure 1-1 includes sync generators, a passive video switcher, a video distribution amplifier, and a waveform sampler. Cameras must be directed from a remote location, video signals switched, and other control functions accomplished. Aside from the purely visual function, many systems also employ warning devices such as transducers, proximity devices, telemetery, and so on.

An example of CCTV for industrial use is in the remote monitoring of gauges and other indicators. In hospitals, patients under special care may be observed or surgical operations monitored. Traffic may be monitored on bridges, bottlenecks, or toll points. One of the fastest growing fields is crime prevention where the use of CCTV is sometimes referred to as security monitoring or more popularly as video surveillance. This is a form of observation used in crime prevention and dramatically aids detection in such places as banks, parking areas, and high risk areas. It is also used, however, for industrial or traffic control (see Figure 1-2) Video surveillance when used in crime prevention or detection is approached in two ways; by "covert" or by "overt" camera location. In covert systems, cameras are hidden. In overt systems, cameras are placed in obvious locations (see Figure 1-3). The idea of an overt system is that individuals will be deterred from theft when they are aware that they are being watched, as when a policeman on a beat is nearby. The covert system, with hidden cameras, makes it possible to identify and apprehend a perpatrator after a crime is committed. In some cases, both methods are employed together. Another approach is to use a dummy camera in an obvious location with an actual camera located for remote observation.

The simplest form of CCTV for any purpose is in the fixed emplacement of a single camera focused on a single area. The camera is connected by standard 75-ohm coaxial cable to a monitor located at some remote point. If there are several areas to be observed, several cameras may be switched by a simple

Figure 1-1.   A Security Camera System. (Courtesy of Dage-MTI, Inc.)

**Figure 1-2.** A "Zoom" Video Surveillance Camera Located in Times Square, New York City.

**Figure 1-3.**  Overt Camera in Times Square.

arrangement. A more advanced solution would be to automatically switch cameras in succession by electronic means. This is known as "sequential switching." Still another method is to cause a respective camera to switch to the monitor when a door opens or when a person accuates a proximity warning or other device.

## ECONOMICS OF VIDEO SURVEILLANCE

The economics of video surveillance are based upon two factors:

1. The money *saved* by the act of surveillance.
2. The money *earned* by doing something that can be done by no other means.

These two factors are somewhat related and are all-encompassing. Under the heading of saving would be monies *not* lost through thefts or through poor industrial monitoring. Money would be *earned* when the use of CCTV makes possible observations that could not otherwise be made, for example in the use of CCTV to observe processes in inaccessible areas.

The arithmetic of saving is simply arrived at once the extent of loss is known. If, for example, it is known that X dollars are lost yearly through theft or vandalism, and if it costs Y dollars per year to buy, maintain, and operate an effective CCTV system, than the savings would be X minus Y dollars. This, of course, assumes that CCTV will be one hundred percent effective in preventing loss. Considering the billions lost per year, it is easy to see why the market for video surveillance has blossomed.

The cost of video surveillance, of course, will have many variables. A receptionist, for example, may be provided with a simple TV monitor and perform surveillance along with other duties. On the other hand, an installation may consist of rows of monitors and alarm boards manned by uniformed armed personnel (see Figure 1–4). In a hospital installation, a head nurse may be provided with a single monitor, or a special nurse may monitor intensive care with an elaborate console.

Maintenance and replacement (amortization) must be included in the cost. Leasing arrangements may also be investigated. Modern solid state electronic equipment not subject to environmental extremes will operate trouble-free if not abused. On the other hand, mechanical devices such as camera and lens drives do break down and wear out. Finally, equipment modernization, upgrading, and expansion is subject to economic guidelines.

## CAMERA MOUNTING

Cameras may be placed on moveable mounts. The simplest form is known as a "scanner." This is a moving mount propelled by a small motor which causes the camera to scan an area from side to side. The angle of scan may be adjustable up to 350 degrees. When action is continuous it is known as *auto-scan*. The scanner may be designed so that continuous scanning may be stopped and the camera moved right or left by switches located at the monitoring center.

When a camera is moved in a horizontal direction this is called *panning*. Movement in a vertical direction is called *tilt*. A mounting

Figure 1-4.   Public Area Video Control Center.

device capable of moving a camera on both horizontal and vertical axis is known as a pan and tilt drive (see Figure 1-5). In its most versatile form, a pan and tilt drive may also be equipped for continuous scan, known as auto-scan. With a pan and tilt drive capable of auto-scan, the observer at the monitor is able to stop the motion and maneuver the camera when an object of interest is observed.

## LENSES AND NIGHT VIEWING

The lens employed on the TV camera determines the amount and distance of the scene that can be observed and the minimum amount of light necessary for a given camera. A lens can give

closeups (telephoto) or it can observe a wide angle. A zoom lens is adjustable for both conditions. A lens can also be made to adjust for different conditions from a remote location by using small electric motors.

Night viewing systems may be "passive" or *active*. Passive night viewing aids amplify the weak reflected light of stars, moon, or dim lights to give reasonable night vision. These aids employ an optical system coupled with an electronic image intensifier. The scene is focused onto the image intensifier and the small amount of reflected light is converted into electrical energy, amplified, then reconverted into visible light.

The intensified image is produced on a miniature TV type display which may be viewed directly or by means of suitable coupling, or it may be viewed on a large size monitor.

Because passive systems rely on moon or starlight they cannot "see" in total darkness. Under these conditions we must supply light and yet prevent those under surveillance from observing the observers.

Active night viewing aids incorporate an infra-red source. Reflected infra-red light is visible to certain TV cameras. The scene under surveillance appears on the monitor, although highlights and surfaces do not appear the same as with ordinary light.

## ENVIRONMENT PROTECTION

The camera of a CCTV system may be required to operate in many environments. For example, an environment may be prohibitive for human observation, such as an area of extreme heat. This would require a cooled camera housing. Cameras placed out of doors must be protected from the elements, and so on. A variety of housing devices are in use for any purpose to which CCTV may be put. These may be classified according to environmental conditions:

- dust proof
- weatherproof
- extreme cold
- extreme heat
- explosion proof

- theft and tamper proof
- high voltage isolation
- nuclear, x-ray or electromagnetic radiation

A housing may protect from one or more of the foregoing. If the camera is required to pan and tilt, then housing weight is a factor. Housings usually provide a small window for the camera lens. If outdoors, provision is usually made for washing and wiping, or at least wiping rain from the window, as well as defrosting or defogging.

## CONCEALMENT

Another form of housing is one required only for concealment in covert surveillance. The housing may be made to resemble a light fixture, grillwork, or other common item. In this case, the lens "sees" through the grillwork or a one way mirror. Dummy cameras in plain sight are used on occasion, either as a deterrent, or as an adjunct to covert cameras. A dummy camera or even an unattended or inoperative camera may even be mounted on an operating scanner as a deterrent. When properly installed, such a device would be difficult to detect from an operative system without close inspection.

# 2

# *Video Security Analysis*

This chapter discusses video security based upon a theoretical approach to security system design. Alarms and camera systems may be combined in an almost infinite number of ways and at a great range of cost. Yet there is an optimum approach to any given situation. To find this approach, it is advisable to begin with a theoretical analysis.

It is often noted that the mere presence of a security measure is a deterrent. Uniformed guards, obvious cameras and warning signs are examples. This may offer psychological protection, but it is not protection against a determined or clever intrusion. A more subtle psychological approach would be to display a camera, even a dummy camera that is in panning motion, with a hidden actual camera. Such arrangements may be all that is required to deter shoplifting and minor misdemeanors.

On the other hand, a determined perpretrator is not deterred by the mere presence of protective devices. Likewise, armed penetration, crowd action, or acts with purported political motivation require that a video surveillance installation be designed from a different set of principles. In this case the purpose of surveillance is to reinforce the ability of protective personnel to take action. This is the ultimate protective function. It takes place after forceable intrusion occurs and physical violence becomes imminent. It may require that the security plan includes the operation of specialized entrance and egress points when specified conditions develop. Certain cameras and alarms may be activated at this time and the surveillance system may be programmed in stages to cover contingencies.

## SURVEILLANCE VS. ALARM

There is a difference between *surveillance* and *alarms*. An alarm indicates a condition. Surveillance, whether through human eyes or a camera, encompasses much more than the indication of a condition. It is virtually impossible to provide absolute surveillance, however, for physical and economic reasons. Alarm devices may therefore be used to initiate surveillance or the dispatch of a patrol.

Closed circuit television as a surveillance device is very effective. Nevertheless, the camera is not as effective as a trained security guard. The theoretical effectiveness of video security can be analyzed by comparing its capabilities and limitations with a trained, armed, security guard or a law officer that is in place. Obviously, the camera has no powers of restraint or apprehension. It cannot rattle a door or detect gases, and it cannot hear. Nor can it approach and peer under concealment. The camera is a single eye—it lacks depth perception except by relative size and location. The camera does not carry a flashlight and must depend upon existing light. It operates from a fixed location.

When comparing alarm devices to video surveillance, however, an advantage of video surveillance appears: when a camera fails, the failure is obvious. When many types of alarm systems become inoperative, the failure can go unnoticed. A greater level of security can often be obtained when video surveillance is used in conjunction with alarm devices. Alarm devices add other senses to surveillance. For example, a camera output may be switched to a master monitor when triggered by vibration, proximity, and so on. Moreover, it may simply be too expensive to provide complete video coverage, just as in the case of human watchmen. The best approach is to analyze each zone of protection and calculate the most economical system under budget restraints.

An important feature of video surveillance is that it can provide a means of individual identification, and also a permanent means of identification when a video recorder is used. Camera surveillance may be bypassed by a sophisticated intruder, just as alarms can also be defeated. A mounted camera reveals its field of view. A camera that is continuously scanning can be bypassed between scans, and a visible, controlled pan and tilt device displays only its field of view. It can still reveal signs of intrusion such as a damaged fence or open door, however.

It is now possible to control a complete security system by computer program, using microprocessors. In this process, certain cameras may be switched on or off, alarms may initiate action, and locks may be released or secured. The method is well within the state of the art, and only requires commercial recognition to reduce the cost. The system may be programmed in a variety of ways, and possibilities are almost unlimited.

## ZONES OF PROTECTION

The basic consideration in security analysis is to define and localize the entity or asset that requires protection. Although it may be assumed that an object or asset requires special protection from human action or activity, there already may have been measures taken for environmental protection such as fire or other catastrophies. These will already be in place and will provide a certain degree of security. The most obvious form of environmental protection is the building itself. The objective of this security analysis is to determine the level of protection which already exists and how to best complement it with various security systems, including CCTV.

We may consider that protection originates from the point occupied by the asset to be protected and radiates outward through five *zones of protection*. Each zone represents a potential layer of security. Figure 2-1 shows these zones. The outermost zone is known as the perimeter. It surrounds the area that marks the beginning of security measures. Normally, it may be a fence, yet it could be only a defined property line. Even lacking a fence, the presence of a "No Trespassing" sign would mark the beginning

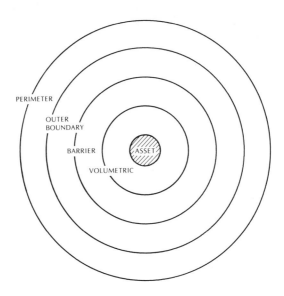

**Figure 2-1.** Zones of Protection.

of the security system. Also lacking a fence, the perimeter may be secured by patrol.

The next zone is the outer boundary. Once having penetrated the perimeter, human presence now exists in this area. The third zone is a barrier or structure. The structure serves as environmental protection from human entry. Then, when the structural barrier has been penetrated, there is human presence within the entire volume of the structure. This is the fourth zone of penetration. The final zone of protection surrounds the asset or entity itself which needs to be guarded against theft, tampering, or mere disturbance.

Zones of protection are theoretical. They apply generally, and may be assigned to all cases. Each zone may be dealt with depending upon its individual characteristics. When a practical case is analyzed, however, we first discover that certain zones cannot be defined, and, secondly, that the zones appear and disappear during day-to-day operations or between night and day. An example would be a public building containing, say, an art object to be secured from theft or vandalism. Here there would never be a perimeter zone, and possibly the exterior zone must be dropped from consideration. During the day, public egress is extended, thus eliminating the penetration and volumetric zones. The proximity zone may be protected in daytime by surveillance. During the night, both structural and volumetric zones materialize, and now three zones—structural, volumetric, and proximity—thus appear. Each zone is dealt with based upon physical and economic considerations. Intrusion occurring in each zone must be separately analyzed. For example, if the outer zone contains flammable objects, materials, or vegetation, the possibility of arson or accidental fire in this zone should be considered. Security devices are chosen according to the capabilities or characteristics of each. Alarm devices and other means are in themselves a broad subject.* All of these devices are a *substitute* for surveillance. Some may be more effective than others, and most may be much less expensive than surveillance in practical applications. Continuous surveillance usually can be claimed as the most effective of all protection.

## VIDEO SURVEILLANCE ANALYSIS

Video surveillance may be classified under the broad headings of *point* surveillance, *area* surveillance, and *volumetric* surveillance. Point surveillance may involve a teller's window or an egress point such as a door. An example of area surveillance would be a warehouse floor. Volumetric surveillance has a three-dimensional quality. That is, surveillance is extended to various levels of interest.

In each case, the first criteria is the field of view of the camera. Point surveillance requires only a narrow field of view. A fixed camera may be used for area surveillance if the field of view is adequate for the given situation. Volumetric surveillance sets up greater requirements and may be obtained only through the use of pan and tilt controls. A remotely controlled zoom lens may be necessary in certain cases of area and volumetric surveillance.

The field of view of a camera is in the form of a cone with the camera lens at the apex. If the camera is capable of being panned,

---

*See Robert L. Barnard, INTRUSION DETECTION SYSTEMS, Woburn, Mass.: Butterworths, 1981 for a full discussion of the various alarm systems.

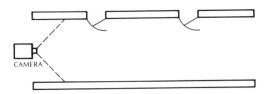

**Figure 2-2.** Lateral Surveillance (Fixed Camera).

the apex angle is increased. Figure 2-2 shows how a camera may be located to cover a field of interest. Note that each has advantages and disadvantages when applied to a given situation. A camera may be located for either lateral or frontal surveillance. In the case of a hall, for example, only lateral surveillance is possible. Figure 2-3 shows how a square area may be covered by two cameras at opposite corners. Certain portions are covered by both cameras, whereas each camera covers the blind spot of the other. Figure 2-4 demonstrates the volumetric class of surveillance.

## MONITORING CRITERIA

In the final analysis, it is the ability of the monitoring operator to perform surveillance that determines effectiveness. This requires analysis of the operator's duties along with the monitoring based upon the following interrelated criteria (assuming that all cameras are at an optimum location):

- number of monitors
- time required to effectively observe each monitor
- scene detail
- events to be observed
- orientation

There is a limit to the number of cameras that can be effectively monitored. On the other hand, the situation may require only a cursory glance at certain monitors. Thus, the time spent in monitoring a scene determines the detail that can be observed. Cameras that are remotely controlled require the full attention of the operator while the monitor is observed. If the operator has other duties, such as making entries, or if monitoring is only a

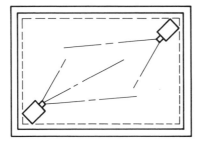

**Figure 2-3.** Area Surveillance (Panned Camera to Cover Entire Area).

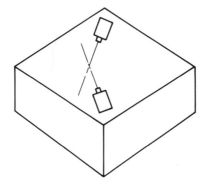

**Figure 2-4.** Volumetric Surveillance (Cameras Equipped for Both Pan and Tilt).

secondary assignment, it becomes less effective. For example, a receptionist may be assigned a secondary duty to monitor a video security installation, as opposed to a security operator with a full time assignment. The details of a scene and the events to be monitored are additional criteria.

The orientation of the observer relative to a camera scene is important. The operation must be able to relate what is seen on the monitor with actual physical objects. Monitor operators should be required to actually visit the area and become familiar with it. The individual should stand at the camera, observe the scene and become acquainted with blind spots, objects, and surveillance requirements. Practice runs may be conducted to determine the ability to recognize events, individuals, and other activity. Practice efforts should be made to defeat the camera. When the system is in operation, security checks may be made with simulated intrusions or other methods.

Studies* have indicated that the performance of monitor personnel decreases under conditions of high temperature and humidity. This suggests that the monitor center should be air conditioned. The best monitoring personnel are women in the mid-20 year-old range.

Design of the monitor console begins with the selection of monitor size. In general, a 9-inch monitor is considered the smallest that will allow the operator to recognize features. Two of these monitors will fit into the standard 19-inch rack employed with electronic equipment. One method is to place a larger monitor at the console center and switch scenes of interest to this monitor as required. Smaller monitors may then be peripherally placed.

The maximum vertical viewing angle between observer and monitor is 30 degrees. A maximum horizontal viewing angle of 45 degrees in both directions is necessary to reduce distortion. Table 2-1 gives the suggested maximum and minimum viewing distances for different size monitors. These criteria suggest that a semi circular monitor console will be optimum. An observer's station should not consist of more than 15 monitors.

When sequential switching is employed it is generally conceded that a maximum of four scenes must be set. More than this saturates

Table 2-1.  Maximum and Minimum Monitor Viewing Distances

| Tube size (inches) | Maximum viewing distance (feet) | Minimum viewing distance (feet) |
|---|---|---|
| 9 | 7.0 | 3.0 |
| 12 | 10.0 | 3.25 |
| 14 | 12.0 | 3.6 |
| 17 | 14.0 | 3.75 |
| 19 | 17.0 | 3.85 |
| 21 | 19.0 | 4.85 |
| 23 | 19.5 | 5.0 |

*B.O. Bergum, "Vigilance, A Guide to Improve Performance," Bulletin 10, October, 1963, Human Resources Research Office.

the observer with information. Increasing the dwell time may relieve this requirement in certain cases. As a general rule, auto-panning cameras should not be used with sequential switching. Many arrangements of manual switching are possible. The output of certain cameras may be terminated without a monitor until the scene is desired. An operator may initiate a microprocessor switching program to meet certain conditions, or a manual procedure may be set up to take the place of a watchman's rounds.

## SUPERVISION

Once an optimum system is installed and operating, control room discipline determines its effectiveness. Observing a monitor can become a boring chore. When sufficient personnel are available, shifts may be set up, with rest periods provided. Unobtrusive random supervisory observation is a good practice. Check lists and standard operating procedures should be enforced.

A monitor station may be operated by a young woman as a primary duty, or by armed persons with the additional duty of physical surveillance, investigation, and aprehension. This will determine operating procedure.

The control room and especially the consoles should be kept clear of impediments. The control room should be a place of business rather than a gathering place. Monitors must be adjusted for proper contrast and focus with the maximum shades of gray for an individual scene, although certain scenes may require special contrast settings. Dwell times for sequential switching must be adjusted for optimum viewing.

Monitor personnel will appreciate sharing the responsibility of maintaining security effectiveness, and boredom can be relieved by encouraging personnel to make periodic intrusion tests and visit individual scenes to check camera operation. Personnel may also be encouraged to become acquainted with individual scene characteristics and suggest changes or improvements. If changes occur in the characteristics and locations of zones of protection, such as lighting conditions, physical changes, etc., monitoring personnel may be the first to notice.

Other than system maintenance, control room efficiency and economy sets the limits on the effectiveness of an installation. Much of this depends upon good supervision and choice of personnel.

# 3

# *Camera Housings*

It was pointed out in Chapter 1 that remote cameras must often be protected by housings. A housing should be no larger or heavier than necessary to provide the intended function. It should be capable of convenient mounting with adequate provisions for safe wiring and finished to prevent rust or corrosion. The camera aperture must not introduce image degradation. A wide choice of housings are available to suit almost every purpose. When specifying a housing, environmental factors are the most important consideration. Tampering and vandalism should also be considered, although a discreet location can help reduce damage. A camera housing may also be required to house transformers or other accessories. This contributes to the generation of heat within the enclosure, which also must be considered. The housing should be of no greater size than necessary for clearance. An important thing to remember is the extra length needed for connectors. About 1½ to 2 inches is helpful for this. Normally, data sheets on cameras from manufacturers do not include these dimensions in illustrations. Small housings are available which are designed for the family of mineaturized 2/3-inch cameras. When measuring for camera size, it is important to allow for the movement of a remote-controlled motorized lens. The lens extends in length when focused to the nearest objects.

## INDOOR ENCLOSURES

When weather is not a factor, the housing chiefly protects against dust and tampering (covert housings will be separately discussed). The typical indoor housing consists of a vented sheet metal structure with a window for the camera (see Figure 3-1). One quarter inch plate glass is usually preferred for the window. The typical indoor housing is not absolutely tamperproof, but is intended to prevent casual interference or at least delay unauthorized access. In its simple form, tampering can be prevented by the use of allen head screws of odd size or so-called tamperproof screws. Either a padlock on a hinged access door or a keyed lock may be used. Guards or authorized maintenance personnel should be issued keys.

In a fixed camera, the housing may be connected to AC line current through knockout holes and conduit. A terminal block or duplex receptacle within the housing serves for connecting camera power and accessories. Video coaxial cable also is connected through these holes, or the manufacturer provides a split plate with grommets. Some indoor housings are provided with a swivel for hang mounting, or specially designed wall or pedestal mounts may be used. The same housing is usually designed for either fixed mounting, scanner mounting, or pan and tilt use.

## DUSTPROOF ENCLOSURES

This family of enclosures is usually built for indoor or outdoor use. A housing which guards against dust will also exclude moisture, thus serving either purpose. The enclosure may be provided with access either through snaps or a series of screw fasteners. Access doors or plates are gasketed and cables enter through a port which is usually caulked.

The chief problem with dustproofing is the maintenance of a clean camera aperture in the presence of microscopic airborne particles. Particles settle and adhere to a surface through electrostatic attraction. In low humidity, wind is capable of building static charges of appreciable magnitude. Other than manual wiping, the method of maintaining a clean window depends upon specific environmental factors, primarily humidity and the nature

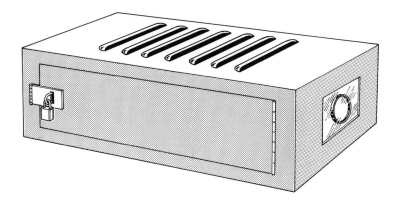

**Figure 3-1.** A Typical Indoor Camera Housing.

of the dust. A remotely controlled window washer (discussed later) is one solution. Other methods may be reached experimentally or through specific experience.

## OUTDOOR ENCLOSURES

A large selection of outdoor enclosures of various sizes are marketed (see Figure 3-2). Factors to be considered in specifying outdoor enclosures are:

● Cooling, ventilation and shielding from sun rays
● Heating
● Size and weight
● Window cleaning, wiping and defrosting
● Mounting of camera
● Access and security

Outdoor enclosures are usually finished in white or shiny aluminum to reflect sun's rays. A sheet metal sunshield may also be added. The sunshield is supported on spacers to allow air to circulate over the top of the enclosure. A blower may also be

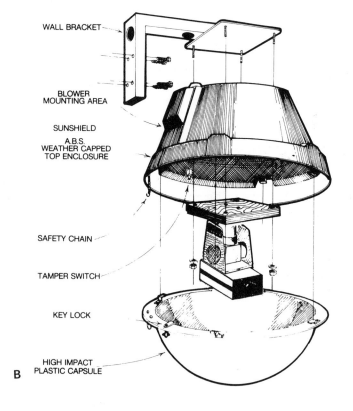

WALL BRACKET

BLOWER
MOUNTING AREA

SUNSHIELD

A.B.S.
WEATHER CAPPED
TOP ENCLOSURE

SAFETY CHAIN

TAMPER SWITCH

KEY LOCK

HIGH IMPACT
PLASTIC CAPSULE

**Figure 3-2.** Two Types of Outdoor Camera Enclosures. (A, Courtesy of Javelin Electronics, Inc. B, Courtesy of Vicon Industries, Inc.)

employed to draw external air in and exhaust warmer air. Another method of ventilation is to employ "chimneys" on the top to exhaust warm air. The air flow should be from bottom to top. Air may also be exhausted across the window. Heating is by thermostatically controlled electrical elements. The usual practice is to maintain temperatures above the freezing level. The window may be defrosted by blowing hot air across it. A cooling blower may be employed for this purpose by combining it with a heating element in such a manner that a common air flow may be used for summer cooling and winter defrosting. Heaters and blowers operate on 115 VAC line current, requiring cabling for this purpose. Remote control is obtained by direct on-off switching or relays operated from the control point. Low voltage relays allow connection by less stringent wiring regulations.

Security may be obtained by locks or bolts. In some instances the main purpose of a housing may be to keep the camera secure. In this case access to the enclosure may only be through concealed bolts and locks and the enclosure constructed of heavy steel. The window may be of bulletproof Lexan or other highly resistant material.

Rain on the camera window will obviously ruin the image. Window wipers almost identical to automobile types are employed for inclement conditions. The wiper turns on and off from the control center (see Figure 3-3) and is made to stop out of the viewing area. Window washers are provided which operate like automobile types. A button at the control center squirts solvent on the window. The wiper is turned on to provide washing action. Solvent is stored in a container and pumped through a nozzle.

Another approach to outdoor protection is to mount the entire pan and tilt unit with camera in a transparent enclosure. The Javelin Model CD29 Outdoor System Housing shown in Figure 3-2 is an example of this. The entire pan and tilt drive is protected from the outdoor environment. This allows the use of a less expensive indoor pan and tilt drive. The unit may be hung from a wall, or the top enclosure may be countersunk in a ceiling. Since, under certain conditions, the dome may fog, a heater and circulation pan are supplied. A tinted dome is available when it is desired to conceal camera movement. This reduces light transmission by about one f-stop.

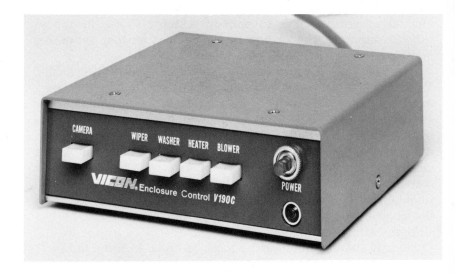

Figure 3-3.  Outdoor Housing Enclosure Control.

## PROTECTION AGAINST EXTREME HEAT

Some situations require observation by CCTV in environments
that are too hot for ordinary cooling methods. Refrigeration
must then be used for cooling. Refrigeration is based upon
the principle of compressing and decompressing air or other
gases. When a given volume of gas is compressed, the heat
energy contained in the gas is now confined to a smaller volume,
resulting in temperature rise. At this point, the compressed gas
can be cooled in some way, for example, by cooling fins on
the air compressor. The cooled compressed gas can now be
piped to the area to be cooled, where it is decompressed by
passing through an orifice. The gas at this point now occupies
a much larger volume, and the opposite takes place, that is, the
larger volume, after decompression contains much less heat
and absorbs heat. Thus, the idea of refrigeration is to compress
and cool the gas in one area, pipe it to the area to be cooled
and decompress it.

One method of cooling camera housings is to obtain air from a standard air compressor. The method is flexible and relatively inexpensive. Certain manufacturers specialize in making cooling orifices and fittings for this method. Sometimes the air supply is already available at the installation. A disadvantage of this method is that ordinary air compressors supply air which contains moisture and a certain amount of oil vapor which causes trouble and must be filtered. Also, air is less efficient as a refrigerant.

More cooling may be obtained by using refrigerant gases which are recycled between the orifice and the compressor in the form of a heat cycle. Ordinary household refrigerators operate on this principle. With any cooling system, the better the insulation, the more complete and efficient the cooling.

## EXPLOSION PROOF HOUSING

Certain environments that cannot be safely observed by other means may be observed with CCTV, if the camera is housed in such a way that no possibility of explosion ignition exists. A housing for this purpose is built according to the National Electrical Code Class I, Division 1, Group C and D; and Class II, Division 1, Group E, F, and G. This housing must be absolutely air tight (see Figure 3-4). It is also possible to pressurize the housing with inert gas, usually nitrogen. A pressure switch may be employed to cut off all power to the housing when pressure falls below a given level. Wires and cable lead into the housing through an explosion proof fitting, and the opening is sealed in with a special sealing compound.

## COVERT ENCLOSURES

The function of a covert enclosure is either to hide or disguise. The enclosure may be disguised as a loudspeaker enclosure, smoke detector, or other innocuous device. The chief problem in the design of covert enclosures is to provide an aperture for the camera which will not degrade the image beyond usefulness. It is possible for a lens to focus through a grillwork placed close to the lens. This will result in softening the image with loss of detail, however. There will also be a loss of light, requiring a larger lens opening, further reducing field depth. When using this approach, greater camera sensitivity is required.

**Figure 3-4.** An Explosive-Atmosphere Must Meet Rigid Specifications. (Courtesy of Vicon Industries, Inc.)

A problem with covert cameras is that camera motion in pan and tilt will reveal the camera. A wide angle lens may be used to eliminate the necessity for camera movement in some cases. It is also possible to house the entire camera with the pan and tilt assembly in a dome treated to act as a one-way mirror (see Figure 3-5). The camera sees through from the inside, the outside appears as a rounded mirror surface. The dome must be carefully designed to prevent distortion and keep light loss to a minimum. Usually about two f-stops are required to compensate for light loss through a mirrored surface even though the silvered surface is kept to as little as one micron thickness of chromium. When locating this type of enclosure it is necessary to eliminate backlighting, which destroys the one-way mirror effect. Also, the camera and other objects within the housing should be finished in dull black to prevent reflections.

One-way mirror housings are sometimes called "discreet" housings. It is possible that the presence of a camera will soon be deduced by observers due to the large bulk and unique appearance of a discreet housing, but an important advantage is that it is impossible to tell where the camera is pointing. There is an opposite method of housing which can be mentioned here that falls into the overt classification. In this method the camera is mounted together with a scanning mechanism in a housing that is deliberately made to give the impression of bulk. This unit, scanning continually, is used for its inhibiting effect. Sometimes a flashing red light may be added for this purpose.

**Figure 3-5.** A "Discreet" One-Way Mirror Housing. (Courtesy of Vicon Industries, Inc.)

# 4

# *Camera Drives*

A camera drive is the device used either to point the camera from a remote point or to cause the camera to scan an area under observation. There are two important considerations in selecting a camera drive; environment and weight of camera assembly. An outdoor camera drive must operate dependably throughout all conditions of winter and summer. A means of braking to prevent movement from wind is a desirable feature.

The majority of camera drives employ a small motor and gear train. Motion is imparted merely by switching the motor on and off from the control point. Reversible motors are used. An AC type of motor employs a separately wired rotor and stator. Reversing either stator or rotor leads in relation to each other will cause the motor to change direction of rotation. This is usually done by running three leads from motor to control point. In permanent locations, wiring regulations require that 115 VAC must be run through conduit or other substantial conductors. This requirement can be eased by using a lower control voltage between the control and camera drive. Some camera drives are therefore designed with 24 AC volt motors. When DC motors are employed, reversing can be accomplished by commutator lead reversal. Permanent magnet DC motors may be reversed merely by main lead reversal.

DC motors with commutators may require RF suppression due to commutator sparking, but they have the advantage of more efficient speed variation with more uniform torque at all speeds. AC induction motors, however, have only one efficient operating speed which is directly related to the energizing frequency. AC motor speed may be varied either by tachometer feedback or silicon diode control.

## OPTICAL SCANNING

Rather than move the entire weight of camera, housing, and various accessories, it is also possible to employ periscope mirrors or prisms to allow the camera to remain stationary while the mirrors move. Such an arrangement is shown schematically in Figure 4-1. The prism is moved in a spindle arrangement around a complete 360 degree circle. A prism or mirror assembly can be made of light materials and caused to move silently with smaller motors. By employing a more complicated design, both pan and

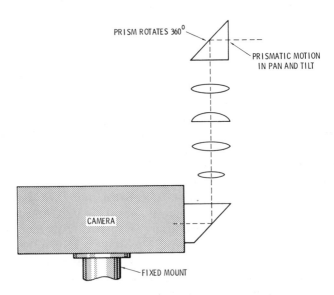

**Figure 4-1.** Rotating Prism and Stationary Camera.

tilt viewing may be permitted. The disadvantage of this system is that, for a combination of reasons, the mechanism must be designed for a specific camera and lens, or the lens must itself have periscopic moving elements. So far little effort has been made to develop a universal design that will provide a choice of camera and lens characteristics. Another advantage of the periscope method is that continuous 360 degree scanning can be obtained, although larger downward tilt angles may be difficult to obtain.

## SCANNERS

Scanners are simple devices designed to move a camera in an arc (see Figure 4-2). They ordinarily consist of a platform mounted on a vertical pintle and driven through a gear powered by a small motor. The speed is usually continuous, although silicon diode circuitry can be employed to vary the speed of an AC motor. A scanner may be connected to a nearby power outlet for continuous operation. Because coaxial cable and wiring extends from the camera, it is not possible to allow a scanner to rotate through the entire circle. Limit stops are used to reverse the driving motor when the scan reaches a certain point. The stops can be preset, allowing the option of scanning in a variable arc which includes only the desired viewing area. Pan speeds vary from three to ten degrees per second, depending upon the manufacturer. Some units can be varied either electrically or by interchanging gears.

Continuous scanning is often called *auto-scan* or *auto-pan*. A problem with this kind of device is the continuous wear introduced by day-to-day operation. When specifying scanners, allowance must be made for replacement due to wear. Other than keeping the mechanism free from foreign matter, moisture, etc., little can be done to prevent ultimate replacement. Units should not operate continuously unless the camera is in use.

Figures 4-3A and 4-3B show the circuitry of scanners which can be controlled from a remote point. Controls are located at the remote point, usually housed together in a metal cabinet. Figure 4-3A shows a scanner operating on 115 VAC. Figure 4-3B shows a 24 VAC scanner. Operation is identical, however 24 VAC is derived from a transformer located in the control. In this manner the wiring extending to the remote scanner carries the lower voltage.

**Figure 4-2.** A "Scanner" Moves the Camera in a Horizontal Direction. (Courtesy of Vicon Industries, Inc.)

The switch marked AUTO-PAN operates when a limit stop is mechanically engaged. This reverses motion as previously described. Power is switched on at the control and push-switches determine camera motion. Right-left switches are interwired so that if both switches are accidentally depressed together, the right switch takes precedence and the left switch is disconnected. A pilot light is provided at the control (see Figure 4-4).

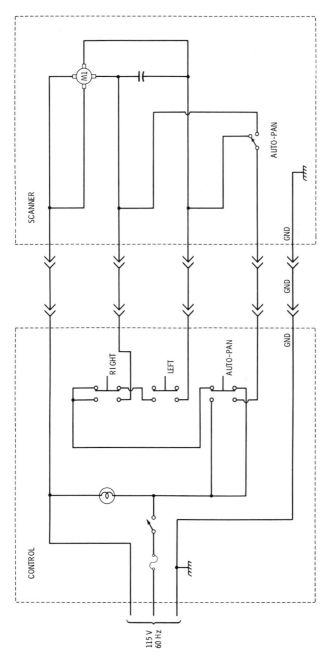

**Figure 4-3A.  Scanner Control Circuitry.  115-Volt AC Scanner.**

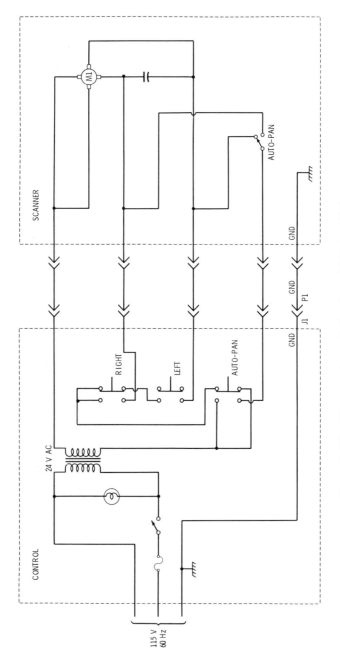

**Figure 4-3B.  Scanner Control Circuitry. 24-Volt AC Scanner.**

Figure 4-4. A Pan and Tilt Drive Control Unit. (Courtesy of Vicon Industries, Inc.)

## PAN AND TILT DRIVES

This device moves the camera on a double axis. For this reason it must be more carefully designed, especially the heavy duty outdoor types. Since design of a scanner or pan and tilt drive depends upon the amount of weight to be swung, this is a convenient method of classification. For camera combinations up to 20 pounds, the classification *light duty* is used; from 20 to about 40 pounds, *medium duty*; and above approximately 40 pounds, *heavy duty* is used (see Figure 4-5). In selecting the unit, therefore, it may be described as *light duty indoor, medium duty outdoor*, and so on.

For discussion purposes, a two axis device may be considered to consist of two rotation centers; the pintle, or lateral column, and the trunion, or tilting column (see Figure 4-6). These columns bear the swung weight. With reference to Figure 4-6A, note that the center of gravity may appear anywhere with relation to the respective pintle and trunion centerline. Ideally, the center of gravity would have to be located at the point in space at the intersection of pintle and trunion centerlines, in order to require minimum drive torque (see Figure 4-7). The designer of a pan and tilt drive may strive to obtain a design as near as possible to this condition. Figure 4-6B shows possible locations for the center of gravity, with the equivalent outline of a pan and tilt drive.

**Figure 4-5.**  A Pan and Tilt Drive Capable of Carrying an 80-Pound Load. (Courtesy of Vicon Industries, Inc.)

Figure 4-8 shows a method of using counterweights to cause the load center of gravity to coincide with the axial center of the drive. In practice, an exact match is not possible due to the variation in load weights of the individual installations. In any case, maintaining balance is important when mounting the camera and accessories. Care must be taken when locating mounting screws, and most pan and tilt drives have several mounting holes for selection.

Load relief may also be obtained from springs. A torsion spring, for example, may be designed to compress upon negative torque and release when the torque becomes positive.

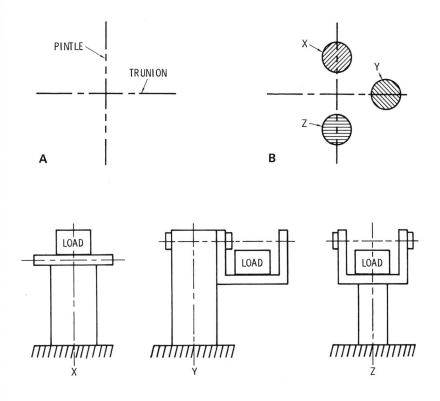

**Figure 4-6.** Center of Gravity Location on Pan and Tilt Drives.

The torque required to swing the load increases with the distance of the center of gravity from the axial center, but becomes negative when the load moves below the trunion centerline. Thus, the gear pass must both resist negative torque and supply a positive torque. Worm gears may be employed to provide both negative and positive torque thereby swinging loads that are off-center. This simplifies construction, although worm gears used in this manner are subject to more wear and, as a class, are less efficient.

Even a perfectly balanced load must be braked against wind load or mount vibration. When worm gears are not used, one method is to use friction brakes that are set against a drum when driving power is off, then lifted by a solenoid when motion begins.

**Figure 4-7.** Medium Duty Outdoor Pan and Tilt Drive. The Load is Located Above the Trunion. (Courtesy of Pelco.)

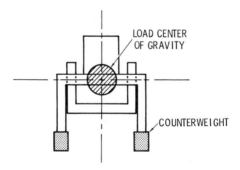

**Figure 4-8.** Counter-Balanced Load to Bring Center of Gravity to Axial Center.

Location of the brake drum at the motor, which is geared down in the order of 250:1, results in negligible reflected torque due to load. When in motion, dynamic braking can be provided by certain motor designs.

The trunion is in motion relative to the pintle, and in auto-pan units, motion is continuous. The tilt driving motor is therefore in motion relative to the fixed portion of the unit. In some designs, both pan and tilt motors move relative to the mounting base. Any wires connecting the sections would thus be in constant flexure. This would result in crystalization and breakage of wire, particularly when in continuous autopan. To eliminate this, slip rings are employed. Stator, rotor, and one common lead are brought through three slip rings. Contact is enhanced by using exotic metals, for example, palladium plated brushes and gold plated slip rings. Limit stops are ordinarily provided which disconnect the motor when limits are reached. By reversing the motor, the unit is driven out of the limit. Auto-pan operation is identical with that of scanners.

A problem encountered with pan and tilt drives is with the cables leading to the camera and to the drive itself. A loop of 3 to 4 feet should be used to help prevent damage to the cables. Some pan and tilt drives are built with a stationary base section to which cable connections are made. In other designs, cable guards must be used to prevent cable snarling when pan and tilt motion takes place. In all cases, the video and camera cables must be tied and arranged to allow free motion.

## CONTROL OF PAN AND TILT DRIVES

A simple method of pan and tilt control is to use the system previously described for scanners. Figure 4-9 shows the schematic of a pan and tilt drive connected to its respective control. This unit uses AC motors. Speed is fixed. The circuit shown furnishes 115 VAC line current directly to the drive, and the drive motors are energized with this voltage. If it is desirable to use low voltage so as to become exempt from house wiring regulations, a 24 volt transformer could be placed in the control, and 24 volt motors could be specified. Another method would be to employ relays.

When DC motors are used, a control such as that shown in Figure 4-10 may be used. This control provides up to 130 VDC which is derived from half-wave diode rectifiers. Pan and tilt speed can be varied by changing transformer taps at the input to the rectifiers.

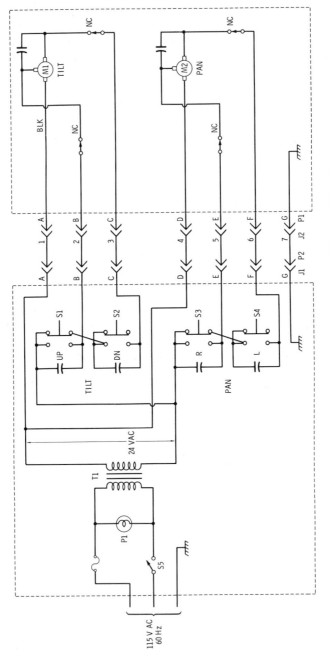

**Figure 4-9.** Schematic Diagram for a 24-Volt AC Pan and Tilt Control.

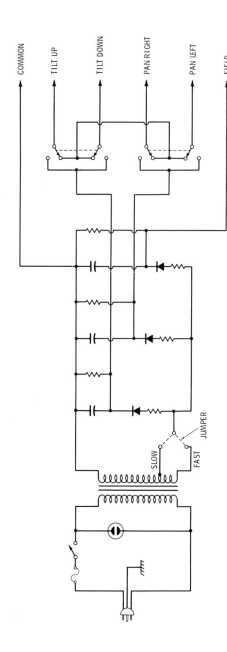

**Figure 4-10.** Schematic Diagram for a Direct-Current Pan and Tilt Control.

# 5

# *Analyzing the Television Signal*

It is now well known that the television image is formed by a thin electron beam that moves across the picture area. This process took years to perfect, however. Primarily, this is because producing the series of images needed to convey a moving image requires millions of small electrical charges per second that must be produced, transmitted, and reproduced by electron beams. After a system was perfected, it was necessary to standardize it so that the products of all manufacturers would be interchangeable.

A television signal is formed by a process of *linear scanning*, that is, building the picture from elements by a succession of scanning lines that move across the picture area. Motion across the picture area is very rapid, but it is nevertheless a uniform motion. The brightness varies at each point to reproduce the image. Our eyes retain this instantaneous amount of light from the scanning process. Additional pictures, called *frames*, are scanned in rapid succession, conveying the impression of motion.

## SCANNING THE PICTURE AREA

The ratio of width and height of the rectangular picture area is the *aspect ratio*. The standard picture is four units wide and three units high. The total number of scanning lines per picture determines the amount of detail that can be conveyed. The standard television picture is based on 525 lines per frame. This number was arrived at because it is an odd number composed of the simple odd factors:

$$3 \times 5 \times 5 \times 7 = 525$$

The number determines how the signal will be divided and subdivided in a synchronous manner. This will be discussed later.

In summary, the standard image in the U.S. is four units wide by three high, and is formed by a beam of electrons that move with a uniform trace across the screen, then very rapidly returns to repeat the process at a point slightly offset from the previous trace. This occurs 525 times for each frame. As we shall see, two "fields" are repeated to form one frame.

## INTERLACED SCANNING

In order to reduce flicker and convey motion, while at the same time keeping the rate of change of the television signal as low as possible (bandwidth reduction), *interlaced scanning* is employed. In this process, the beam makes two trips through the picture area. *Each trip forms a field*. One half, or 262.5 lines scan the picture area to form a field, then the remaining half repeats in between the spaces to form the complete frame. This action is known as *two-field odd-line*, and is illustrated in Figure 5-1.

A complete scan begins at Point A. When the beam completes the luminous portion of the trace, the beam is blanked and a rapid retrace takes place. This horizontal retrace period is very rapid relative to the scanning period. With each retrace, the line is moved downward slightly. When the beam reaches the bottom of the picture area, it is again blanked. The blanked beam then moves rapidly upward in a zigzag pattern until it is at position E. It then begins another *field*, staggered or interlaced with the

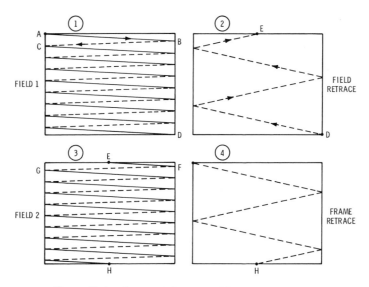

**Figure 5-1.**  Standard Scanning Method (See Text).

previous field. When reaching point H, it is blanked and zigzags back to Point A. This completes one frame. Each field occurs 60 times per second, hence, each complete frame occurs 30 times per second. There are now 262.5 field lines scanning 60 times per second, or one 525 lines 30 times per second. With 525 line periods per frame and 30 frames per second there will be 15,750 line periods per second. Of the 525 lines in each frame, 35 lines are used up while the inactive beam is making the two vertical return excursions. This leaves 490 luminous lines left to form the picture.

At this point, we should examine the subject of picture resolution. Since there are now 490 effective scanning lines, a vertical white line in front of a TV camera would appear broken up into 490 pieces. Since each scanning line has a discrete width, it is obvious that some scene detail will be lost between the lines. As a general rule, about 30 percent of the scene may be lost in this way. Thus, about 343 lines of resolution (0.7 $\times$ 490) can be expected from a standard 525 line system.

Note that this is the resolution that can be expected as a vertical line is broken up. In other words, the figure of 343 has to do with *vertical* resolution. *Horizontal* resolution depends upon how fast the electron beam changes intensity as it traces the image. The maximum speed at which horizontal changes occur depends upon frequency response or the bandwidth of the television signal. In other words, although vertical resolution is determined by the number of scanning lines chosen (525 by U.S. standard), horizontal resolution depends upon the electrical performance of the individual system. In general, if a system is capable of transmitting up to 4 MHz, a horizontal resolution of 340 can be attained with properly operating equipment.

A practical figure of merit for a television system would be the total number of picture elements, or dots, which may be reproduced within the picture area. This would be the product of the vertical and horizontal resolution. A value of 165,000 picture elements would be a figure for a good 525 line CCTV system.


## SYNCHRONIZING THE SCAN

It is now necessary to examine the means by which scanning is caused to take place. The complete black and white television signal must be able to do two things; it must be able to convey the impulses to cause the beam at the receiver to vary from light to dark, and it must mark the points in time at which the scan is to be caused to start and stop, (which includes the times that the beam must be blanked out to allow for retrace and sync). The video portion of the signal itself appears on a scope trace as a jumble of lines, as the beam "paints" in the image. The synchronizing pulses, however, are distinct in both time and amplitude.

Standard RETMA television sync waveforms are shown in Figure 5-2. The basic frequency of the sync is 60 Hz. The reason for this is that by the time television standards were adapted, the powerline frequencies in North America had already been standardized to an accurate 60 Hz and were being held to Bureau of Standards frequencies within reasonably close limits. This permits synchronization in simple systems to line current frequencies, but also prevents *heterodyning* or a flickering beat which could occur when line and synchronizing frequencies are not the same.

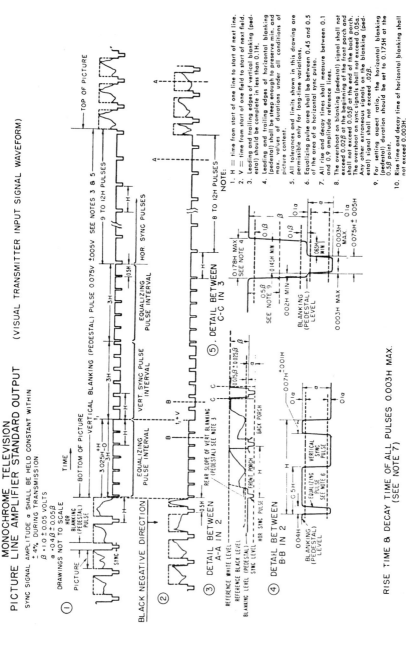

**Figure 5-2.** Standard Television Sync Waveforms. (Courtesy of Electronic Industries Assn.)

Synchronizing standards are based upon the duration of one scanning line, or the reciprocal of the scanning frequency (1/15,750 in the case of the U.S. Standard). The symbol H is used to represent this duration of one scanning line, and represents the time interval between any one point on a horizontal pulse and the identical point on the following horizontal pulse. All synchronizing parameters, such as pulse rates and durations, can be expressed in terms of H time. Referring to Figure 5-2, note that the active part of a scanning line that is actually developing a video signal will be H minus blanking pulse duration. Also, the time during an entire vertical field that actual video information is transmitted will be equal to field time minus vertical blanking time as well as the horizontal blanking time which we have already considered. The waveforms shown in Figure 5-2 are those that would be produced by a television sync generator and fed throughout a system. The synchronizing pulse standards are listed in Table 5-1. If a separate sync generator is not used, such as in a simple system, the signals necessary to generate the composite signal must be developed within the camera itself. More will be discussed about this with the subject of cameras and monitors.

The standard composite television signal representing one horizontal scanning line is shown in Figure 5-3. Note that the standard applies to amplitude as well as time (pulse shape). This is the critical waveform which must be dealt with in CCTV. It must be transmitted and received with reasonable fidelity in any system. The highest frequency component is developed when the sync portion of the signal rises and falls. Other high frequency components are formed when the signal changes rapidly from light to dark picture levels. As mentioned previously, about 4 MHz components must be passed for good picture reproduction.

We will return to the subject of synchronization later when cameras and entire systems are discussed. The important thing to note at this point is that in closed-circuit television, the burden of signal transmission, hence, signal quality is upon the user, rather than the FCC regulated commercial broadcaster. Cost of installation, system complexity, and picture quality are trade-offs which must be settled by the user of CCTV.

Table 5-1. Synchronizing Pulse Standards

| Pulse identity | Pulse rate (freq in pps) | Duration (H-63.492 μs) (V-16,677 μs) | Duration in microseconds | | | Repetition rate (Hz) |
|---|---|---|---|---|---|---|
| | | | Min | Nominal | Max | |
| H. Sync | 15,750 | .075H ± .005H | 4.44 | 4.76 | 5.08 | 15,750 |
| H. Blanking | 15,750 | .165H to .178H | 10.48 | (10.89)* | 11.3 | 15,750 |
| Equalizing | 31,500 | .04H† | 2.00‡ | (2.27)* | 2.54‡ | 60 |
| V. Sync | 31,500 | .43HH to 01H | 26.67 | 27.3 | 27.9 | 60 |
| V. Serration | 31,500 | .07H ± .01H | 3.81 | 4.44 | 5.08 | 60 |
| V. Blanking | 60 | .075V ± .005H | 1167 | 1250 | 1333 | 60 |
| H. Driving | 15,750 | .1H ± .005H | 6.03 | 6.35 | 6.67 | 15,750 |
| V. Driving | 60 | .04 ± .006V | 567 | 667 | 767 | 60 |

*Midpoint of range

†To be .45 to .50 area of H. Sync

‡Calculated values

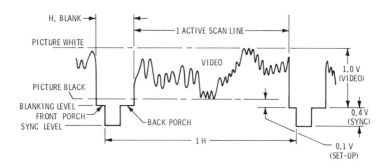

**Figure 5-3.** Standard Composite Television Signal.

## NARROW BANDWIDTH TELEVISION

We have seen that the resolution of a single TV frame is a function of time. Vertical resolution depends upon the total number of lines per second. Horizontal resolution is a function of the rate of change of the video impulses. We would attain greater resolution by cutting down the frame rate, but this would cause image motion to appear jerky, as in old-time movies. Even then, several megahertz of bandwidth would be required. In the extreme, we could dispense with motion pictures. This would require only a narrow bandwidth in the audio range. Such a signal could be sent over a telephone circuit.

The advantage of being able to send a picture over simple wire or radio is cost. In fact, the cost of one or two miles of coaxial cable may be more than that of a high grade camera and monitor.

Sending pictures at very low frame rates is a trade-off between the resolution desired for a given bandwidth. Table 5-2 shows the relationship between the independent variables of bandwidth and frame time and horizontal and vertical resolution. Note that if we are willing to wait one half minute for a frame to be formed in some way, we get rather excellent resolution in the order of 700 lines, or we may form an image with 500 lines of resolution in one dimension and 960 in the other. And this can be done without exceeding an audio frequency of up to 8 kHz.

**Table 5-2.** Relationship Between Bandwidth, Frame Time and Resolution in Slow Scan Television Systems

| Effective bandwidth | Elements/s | Frame time | Resolution in pixels |
|---|---|---|---|
| 1 kHz | 2000 | 30.00 s | 250 × 240 |
| 1 kHz | 2000 | 1.00 min | 250 × 480 |
| 1 kHz | 2000 | 2.00 min | 500 × 480 |
| 2 kHz | 4000 | 30.00 s | 250 × 480 |
| 2 kHz | 4000 | 1.00 min | 500 × 480 |
| 2 kHz | 4000 | 2.00 min | 500 × 960 |
| 4 kHz | 8000 | 7.50 s | 250 × 240 |
| 4 kHz | 8000 | 15.00 s | 250 × 480 |
| 4 kHz | 8000 | 30.00 s | 500 × 480 |
| 4 kHz | 8000 | 1.00 min | 500 × 960 |
| 8 kHz | 16000 | 3.75 s | 250 × 240 |
| 8 kHz | 16000 | 7.50 s | 250 × 480 |
| 8 kHz | 16000 | 15.00 s | 500 × 480 |
| 8 kHz | 16000 | 30.00 s | 500 × 960 |

## METHODS OF ACHIEVING NARROW BANDWIDTH

The total scanning process may simply be slowed down in order to produce an image using only a narrow bandwidth. The difficulty of this is that on an ordinary cathode ray screen the eye would see only a moving dot with varying brightness and no image would be apparent. A screen of long persistence would solve this problem. Using a persistent phosphor, we would see an image which would gradually fade away unless renewed by another scan. The effect of this is the same as on the familiar radar PPI scope. Another method of receiving slow scan would be to expose photographic material to the moving spot in a darkened enclosure. Upon chemical development, a permanent image would appear.

For marketing and economic reasons, rather than use special slow scan cameras and monitors, it would be advantageous to use standard 525 line format equipment. One way would be to use a *scan-converter*. This is a device using a double-ended tube in which an image may be traced on a screen in slow scan format, then picked up by a separate beam which is moving at standard rate, or vice versa. Narrow Band TV using scan conversion has been used in space missions, but the process is complicated and expensive.

A more recent approach has been developed by Colorado Video, Inc. of Boulder, Colorado, in which image information in standard format is stored on a magnetic disc. Disc information is then sampled at a slow rate. At the receiving end, the slow information is again recorded and picked up at the higher standard rate for display on a conventional monitor.

This system operates by sampling the standard TV signal once during each 63.5 micro-second horizontal scanning period with a sample and hold circuit. This amounts to scanning the standard TV picture vertically at a rate of 1/30 s/frame. Thus 525 lines are sampled each 1/30 second. A slow horizontal scan is thus superimposed on the vertical scan so that the picture is scanned left to right at a rate of 8 s/scan (seconds per scan). At the end of 8 seconds, 240 vertical scans of 525 lines each have been made resulting is 126,000 samples/frame. Since each H line has been sampled 240 times/frame, the maximum time represented by one sample is 63.5/240 $\mu$s. or 264 nanoseconds. This allows the sample and hold circuit 63.5 microseconds to transmit this sample of information. Time has therefore been expanded by a factor of 240 with a resultant decrease of transmission bandwidth.

At the receiving end of the system, a new sliding pulse stream is generated. These pulses are pulse-width modulated by the incoming narrow-bandwidth signal and recorded on a magnetic disc which is revolving at 1800 r/min or 30 r/s. With each revolution of the disk, 525 pulses, one for each line in the frame, are recorded. The first 525 pulses correspond to the leftmost samples. The pulses in the second group, displaced slightly from those in the first group, constitute the next vertical scan; the frame is thus built up left to right as the disk continues to revolve.

At the end of 8 seconds, the disk has revolved 240 times and has recorded the 126,000 samples, or one complete frame. The disk is played back in real time and displayed on a standard TV monitor. After a frame has been recorded, the transmitter may be taken off the line until a different frame is to be sent. A received frame may be displayed as long as desired from the disk recording.

The narrow-band system of signals such as described can rarely be sent over lines in "raw" form. Usually it is used to modulate a carrier and then demodulated at the receiving end. The subject of modulation will be more thoroughly discussed in the chapter entitled "Control of Video Accessories." A complete narrow band TV system is shown in Figure 5-4. The device for converting from standard TV format to narrow band is called a *video compressor*. At the receiving end a *Video Expander* is used to convert back to standard format for conventional monitor display.

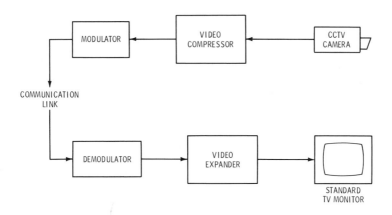

Figure 5-4.   A Narrow-Band Television System.

# 6

# *Transmitting the Video Signal*

Closed Circuit Television implies that a video image is transmitted over a closed circuit rather than broadcast. The circuit in this case must be capable of satisfactorily carrying a video signal. Ideally, this would mean that a current of 30 Hz must be able to flow on the line in the same manner as a 4 MHz current and vice versa. Although this ideal is not attained in practical systems, certain tolerances must be met for a given system to operate properly.

## TRANSMISSION LINES

To understand the requirements of video transmission, it is necessary to review the rudiments of transmission line theory. Alternating current such as that of a video signal does not flow uniformly through a wire as does direct current. This is because any wire also exhibits properties of inductance and capacitance. When the wire increases over a few feet in length the effect of inductance and capacitance becomes important. Moreover, unless the wire is enclosed in a shield, the wire may also act as an antenna, radiating a portion of energy.

It may be noted in passing that the nation's power transmission lines radiate hundreds of thousands of watts of RF energy at 60 Hz. Small amounts of this energy may be picked up wherever an unshielded wire is stretched so that it acts as a receiving antenna. When a wire is close to an AC line carrying large amounts of current, there is also inductive pickup. This manifests itself as "hum," which can be a source of trouble and will be discussed later.

A simple transmission line consists of two conductors separated by a fixed distance and insulated from each other. The two conductors thus running parallel to each other will appear as a long capacitor. To put it another way, for every foot of distance along the two wires there will be a fixed value of capacitance appearing in shunt. Furthermore, the wires exhibit a series inductive effect upon each other. Hence, for every running foot of transmission line, there will be a fixed value of inductance. Figure 6-1 shows how the line appears in circuit form. Engineers consider a transmission line as composed of an infinite number of series inductances and shunt capacitors with a certain value per running foot of line.

A simple line such as described, however, must operate away from all other objects. If another conducting object is nearby, the circuit will be affected by it. To make the circuit act as a pure transmission line, one conductor is enclosed within the other and separated by an insulating (dielectric) material (see Figure 6-2). This now becomes a coaxial cable, or *coax*. The outside conductor isolates the inside conductor which has the effect of isolating the entire line from external objects. The outside conductor is usually placed at ground potential, but is also covered with insulation in most coaxial cable used in CCTV.

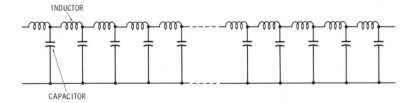

**Figure 6-1.** Transmission Line Equivalent Circuit.

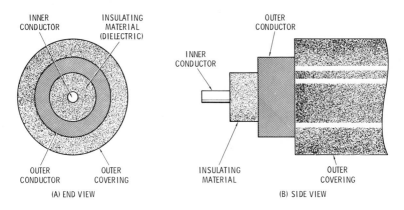

Figure 6-2. Coaxial Cable.

## ATTENUATION IN COAXIAL CABLES

Returning to Figure 6-1, we note that energy may be alternately stored in either the inductive or capacitive component. If a surge of energy (such as a pulse) is applied to the line, the pulse proceeds down the line by alternately storing its energy in the inductive and capacitive component. The pulse actually progresses at a speed slower than the speed of light. It is interesting to note, for example that it took several seconds for a telegraph impulse to pass across the Atlantic on the old trans-Atlantic cables.

Because the cable consists of interacting elements of capacity and inductance, it is frequency sensitive. A glance at Figure 6-1 shows that a transmission line is actually a low pass filter. Thus, higher frequencies may be cut off almost entirely. Low frequencies pass more easily, but at less velocity. We now have two considerations, variations in velocity and variations in attenuation. Attenuation characteristics of various types of coax are shown in Figure 6-3.

A problem in transmitting video over long distances is the varying attenuation of the different frequencies of the video spectrum. For example, the attenuation of a 10-MHz signal is approximately three times greater than that of a 1-MHz signal when using RG-11/U cable. With a 3000-foot cable the 5-MHz portion of a video signal will be attenuated to about one-fourth

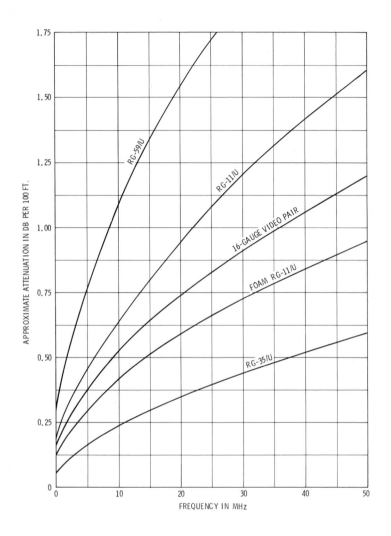

**Figure 6-3.** Coaxial Cable Attenuation Characteristics.

its original level (12 DB), whereas the 1-MHz portion of the signal will be attenuated to only half its original level (6 DB). At frequencies below 500 KHz, attenuation is negligible, causing the higher-frequency portions to be lost. High-frequency deterioration of 6 DB at 8-MHz is noticeable in a quality system. On a standard television receiver, losses in the order of 10 DB would be noticeable. This deterioration appears on a monitor as loss of definition and poor contrast. Later in the text, it will be shown that high frequency losses can be recovered somewhat by selective amplification.

## TERMINATION

If a pulse is transmitted along a line that is open at its far end, the pulse will reach the far end and bounce or reflect back to the source. It has been mentioned that there is a certain amount of delay in transmitting the signal. The result of this is a "ghost" picture appearing on the TV monitor. To prevent this as well as other degrading effects, a coaxial line must be terminated at both ends in its *characteristic impedance*.

The characteristic impedance of a coaxial line is determined by the relative diameter of the inner and outer conductors and the characteristics of the dielectric used in separating the two. Actually, a cable does not have the same characteristic impedance to all frequencies. For example, a cable with an impedance of 75 ohms at 5 MHz may show an impedance of 1000 ohms at 60 Hz. Incidentally, the characteristic impedance approaches infinity as frequency nears zero (zero frequency is a direct current). Usually, the terminating impedance specified for a cable is determined for the most troublesome frequencies at which "ghosting" and other degradation may occur.

When longer cable lengths beyond 1000 feet are encountered, more attention must be paid to proper termination. At some point, it may be necessary to employ a so-called *complex termination*. This is an arrangement of resistances and reactances arranged to provide correct termination at various points in the video spectrum. At frequencies below 500 KHz, impedance rises sharply, hence, special circuitry is provided below these points.

It has been pointed out that characteristic impedance is a function of inner and outer cable diameter. No cable can be perfectly manufactured. Hence, slight variations of dimensions along the cable will result in different impedances. The smaller the cable, the more effect a small variation will have. Thus, larger cable tends to exhibit more uniform characteristics and must be used for longer lines.

## BALANCED VIDEO TRANSMISSION

The purpose of a balanced line is to prevent hum and "crosstalk," that is, mutual interference in video circuits. Figure 6-4 shows the arrangement of a balanced transmission line. The line is fed and terminated by center-tapped transformers. The center tap is connected to the shield and provides a common ground return for the two surrounded center conductors. Any interference on the balanced line such as hum or induced noise will thus be cancelled out. The balanced line must be terminated in its characteristic impedance just as the unbalanced line. To do this, the terminating transformers are designed for a specific impedance, or if necessary a complex termination is employed.

Standard forms of coaxial cable have been developed for in–city video distribution. The impedance is standardized at 124 ohms. Attenuation on this cable is about 0.5 Db per 100 feet. Most video security installations do not require balanced

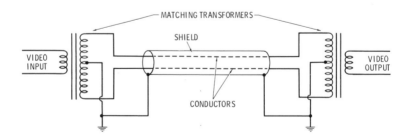

Figure 6-4.  Balanced Transmission Line.

line. Use of the more expensive balanced line is warranted in more sophisticated installations, or where hum and noise cannot be eliminated in any other way.

## FIGURING TRANSMISSION LINE LOSS

A unit of gain or attenuation is the *decibel*. Because the decibel is a logrithmic or exponential unit, and because losses on a transmission line are also exponential, the decibel is a handy method of describing attenuation. The decibel is not a unit of power, however. It is actually the logrithm of the ratio of two amounts of power. When the quantity of decibels carries a negative sign, it is a loss. An amplifier has gain and the amount of gain is expressed in positive decibels. If a signal is applied to a 500 foot line having —1 Db. per 100 feet of attenuation, the signal will be attenuated —5 Db. If an amplifier with +5 Db. of gain is used at the far end, the output will be 0 Db, indicating that the amplified signal at the end is exactly equal to the input signal.

## CCTV TRANSMITTING CABLE

Four types of coaxial cable are used for video transmission (See Table 6-1).

1. 75 ohm unbalanced, indoor
2. 75 ohm unbalanced, outdoor
3. 124 ohm balanced, indoor
4. 124 ohm balanced, outdoor

Cable selected for use in an installation depends upon environment, electrical demands, and economics.

Indoor cable is smaller in diameter using a braided shield which is much more flexible. It can be formed around corners or hidden. Because overall diameter of the cable is smaller, the inner conductor must be proportionately smaller in diameter. A decrease in diameter of the inner conductor causes an increase in attenuation especially at higher frequencies. Small–diameter cable is more susceptible to damage than larger cable. Tight bends, kinking,

**Table 6-1.** Types of Coaxial Cable Used for CCTV

| Use | Type | Designation |
|---|---|---|
| Overhead, duct or patch cables; indoor installation | 75-ohm unbalanced | RG-59/U 8281 |
| | 75-ohm low-loss unbalanced | RG-59B/U Foam RG-35B/U |
| Overhead, or duct; indoor or outdoor installation | 75-ohm unbalanced 75-ohm low-loss unbalanced | RG-11/U RG-11/U Foam |
| Direct earth burial | 75-ohm unbalanced | RG-13A/U |
| Overhead or duct; outdoor installation | 124-ohm balanced | 16 PEVL 16 PSVL V-1-AL VP-1 |
| Indoor installation | 124-ohm balanced | 754E T-43 FVP 219 |
| Patch panels | 124-ohm balanced miniature | T-43M FVP 224 |
| Direct earth burial | 124-ohm balanced | V-1-DSAL VP-PCP-1 |

indentations, or other irregularities cause a change in characteristic impedance at that point.

When indoor cable is used, extremely short bends should be avoided, overall length should be short, and it should not be walked on or subjected to pressure. Indoor cable may be temporarily used outdoors, but it will break down in the outside environment after a short time.

Outdoor coaxial cable is of larger diameter than indoor types and has insulation which is resistant to the elements, physically stronger, and withstands higher voltage breakdown. Because outdoor cable is used for greater distances, it is also designed for less attenuation. Outdoor cable may be buried, suspended from poles, or run along outside structures. Buried cable withstands more mechanical stress and, since underground temperatures are more constant, both characteristic impedance and attenuation remain more constant.

Construction of coaxial cable is shown in Figure 6-5. For indoor cable, the outside conductor is in the form of a tubular braid which lends flexibility. Outdoor cable may use copper wrapping for the ouside lead. More durable cable may use tubing of copper, aluminum, or even lead. The center dielectric may be foam or solid plastic.

## VIDEO CONNECTORS

Connectors used with coaxial cable are designed so that a connection can be made without causing appreciable mismatch. In other words, physical dimensions of the conducting elements of the connector are such that they match the characteristic impedance of the cable on which they are used. Although a connector is designed for a specific impedance, it is not always necessary to have an exact match. For example, a 50 ohm connector may be used on a 75 ohm cable without appreciable problem.

Figure 6-6 shows eleven types of connectors found in use. The two types encountered the most are the BNC, used indoors, and the UHF, which may be used outdoors. These types are used with 75 ohm unbalanced coax and may be considered standard. They are used with the smaller, flexible types of cable. Less flexible types of cable must usually be soldered in place.

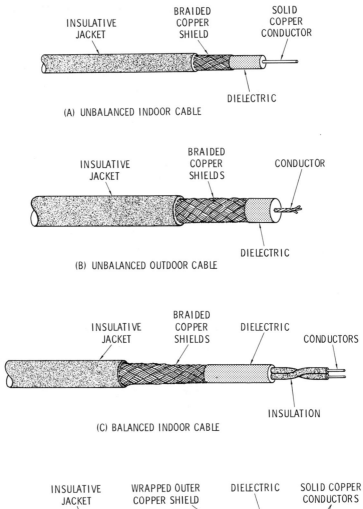

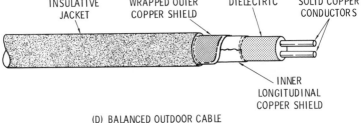

Figure 6-5.   Coaxial Cable Construction.

| CABLE TYPE | DIA | CONNECTOR TYPE* |
|---|---|---|
| UNBALANCED CABLE | | |
| RG-59/U & RG-59B/U | .242 | (A) UG260B/U<br>(A) IPC 84825<br>(B) PL259A or (B) NT 49195<br>with (C) Adapter UG-176/U<br>(D) UG-603/U<br>(D) UG-603A/U |
| 8281 | .304 | (E)<br>(B) PL259A or (B)<br>with (C) Adapter |
| RG-11/U | .405 | (B) PL259A<br>(B) NT 49195<br>(D) UG-94A/U |
| RG-13A/U | .420 | (B) PL-259A<br>(B) NT 49195<br>(D) UG-94A/U<br>(A) UG-959/U |
| RG-35B/U | .945 | (D) UG167B/U |
| BALANCED CABLE | | |
| T-43M | .242 | (E) |
| FVP-224 | .242 | (E) |
| TWC-124-2 | .242 | (F) |
| BL1242 | .242 | |
| 754E | .420 | (G)<br>(H) UG-421B/U |
| T-43 | .420 | (I) UG-102/U |
| FVP-219 | .420 | (J) |
| 5305-8 | .420 | |
| 16PEVL & | .440 | (G) |
| 16P SVL | .440 | (H) UG-421B/U |
| V-1-AL | .440 | (I) PL-295 |
| 16 AWG | .440 | (I) PL-284 |
| VPC-1 | .460 | |
| V-1-D SAL | .600 | (I) UG-1060/U |
| VP-PCP-1 | .600 | (K) IPC-13875 |
| * Letters in circles designate connector illustration. | | |

**Figure 6-6A.** Eleven Connector Types. Cables.

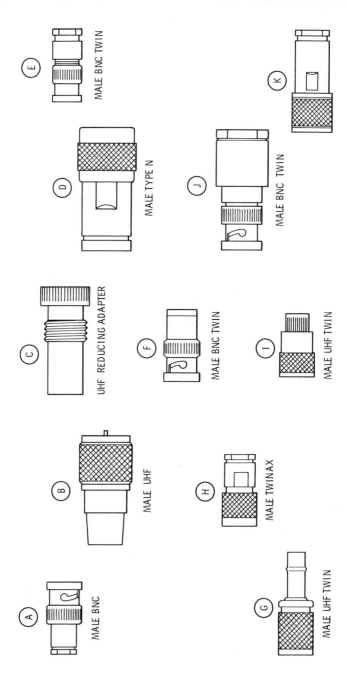

**Figure 6-6B.** Eleven Connector Types. Connectors.

# 7

# Cameras, Monitors and Video Recorders

The first television cameras were necessarily bulky and cumbersome. Moreover, the image tubes lacked sensitivity and required very bright lighting. The last few years have changed all this. Pickup tubes have greatly improved in sensitivity, especially with the introduction of silicon types. Transistors and integrated circuits make lightweight, much less expensive cameras possible. Complete color cameras are now available that can be held in the hand and are provided with pistol grip type handles. All of this has brought CCTV into wider applications.

The function of the CCTV camera is to convert the light image into the video signal. In CCTV systems, this usually means that the camera generates the entire composite signal, including blanking and sync pulses. The function of sync generation can be considered as separate from the image signal. The sync pulse is the independent variable of a television signal. It is the sync signal that "tells" the camera and monitor that a scanning line is about to start and regulates signal parameters.

A synchronizing signal may be provided by a separate unit, or it may be part of the camera. Most cameras may be synchronized either through their own internal sync generator or through an external source. Connectors are provided for separate sync. There are thus two separate systems taking part in forming one television signal. One system deals with the processing of the video portion of the signal, the other the sync. Our discussion of cameras can begin with the circuitry necessary to generate and regulate the video signal, then we will examine the methods in which the sync is generated and mixed with the video to form the composite television signal.

## THE VIDICON TUBE

The development of the vidicon tube has made television possible as we know it. The vidicon tube takes the image as it is created by the lens and reduces it to electrical impulses that ultimately become the television signal described in the last chapter. It is a light–to–electrical signal converter consisting of a glass tube with connecting pins on one end and an optical image surface on the other end (see Figure 7-1).

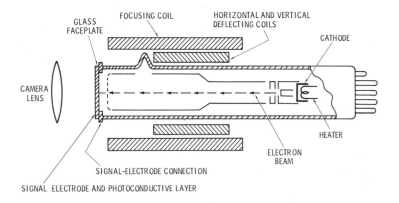

**Figure 7-1.**  Vidicon Tube Cross-Section Drawing.

A source of free electrons is formed by a cathode consisting of certain elements of rare-earth composition which are in turn heated by a filament (thermionic emission). The principle is the same as used in vacuum tubes and cathode ray tubes. Free electrons are accelerated by a positively charged structure, then focused into a narrow beam. This narrow beam, or stream, now collides with the flat image end of the tube which is coated with a photoconductive layer. Electron energy is released at the point at which the beam strikes the photoconductive layer. The amount of energy released is determined by the amount of light falling upon the surface at their point. As seen in the Figure 7-1, a signal electrode is present which now receives a charge that depends exactly upon the energy released by the signal electrode. At this point, a weak current originates which is in proportion to the amount of light.

The image, which has been formed by the lens, consists of varying amounts of light from each point of the surface. The electron beam is now caused to sweep the image area. In areas where there are larger amounts of light, the signal electrode forms a greater current. The image sweeps the area in a scanning pattern described in the last chapter. Hence, a signal is created which may in turn be used to recreate the image. Coils which surround the vidicon generate a magnetic field that deflects the beam in exactly the correct manner to form a standard television signal. The weak current from the signal electrode is amplified as required.

## VIDEO PROCESSING

The signal delivered by the pickup tube is a low amplitude, often only several microvolts. This signal must be amplified to the standard 1-volt peak-to-peak value. In accordance with adopted standards, the signal peak for an entire excursion from black to white for the average scene must equal this standard 1 volt value. This requires that both a black level and a white level must be established for each scene.

The limit to the amount of amplification that can be accomplished depends upon the tiny residual currents, or "noise," that is present in any circuit or electronic device. This "noise" is caused by random molecular disturbances, the passage of electrons, and other sources. Once these small currents exceed the amount of small signal that is to be amplified, then these noise currents, which are also amplified by the same amount, cannot be separated from the desired signal. Field-effect transistors have a high input impedance and can be used for the first (and critical) amplification stage. It is the input to the first amplifier that determines the amount of noise which will be included in the final signal.

Somewhere, either in the preamplifier or amplifier, circuitry is often provided to improve response to either the low or high frequency signal components. A small inductance may be used to peak high frequency transitions of the signal and negative feedback can be used to improve low frequency response. Integrated-circuit (IC) operational amplifiers make good video amplifiers. Certain IC chips are especially designed as video amplifiers.

The human eye has a nonlinear response to light stimulus. It happens that a vidicon tube has a response that closely resembles that of the human eye. Silicon diode pickup tubes have a more linear response, however, and would ordinarily not give a compatible image. The relationship between light stimulus and response is known as *gamma*. It is related to the ability of a system to respond to the shades of gray in the same manner as the human eye. Gamma correction can be made by employing circuitry that will have a reciprocal response, that is, a response in the opposite proportion to video magnitude. One method of gamma correction is to employ the active portion of the resistance/voltage curve of a diode to provide a nonlinear transfer characteristic, wherein the voltage drop of the diode will not be a linear function of current. When video is passed through the circuit, it will be modified by the nonlinearity of the resistance/voltage relationship in a manner that makes the final image compatible with that which would be observed by the human eye.

*Clamping* circuitry is used to establish a DC zero reference level. Clamping is necessary because circuit nonlinearity may cause a signal to ride up or down with reference to the DC level. There are two limiting levels to the video signal: white level and black level. Ideally, the highest signal value should be achieved when scanning the most white portion of a given frame. Black level is the level that will produce a completely black screen. Sync pulses operate in the area below the black level, hence, the screen remains black during sync, including the retrace period. Circuitry may be designed to automatically hold both black and white levels. Summarizing level control circuitry, we can say that circuitry may be required to hold black level, white level, or average level. Each camera designer may include one or all categories depending upon design criteria.

## SYNCHRONIZING

Synchronization is a separate function. As we have previously established, it is the independant variable, a time function, and it tells the system when the horizontal sweep is to begin and when a frame and field begins and ends.

In Chapter 5 where we discussed interlaced scanning, we showed how the horizontal scanning line is returned to exactly one-half line between fields to provide geometrically symetrical scanning. This is in accordance with EIA Standard RS-170 and requires that horizontal and vertical sync must be exactly locked together. In CCTV, however, it is permissible to allow the horizontal scan to run free of the vertical, providing the horizontal scan is within one or two percent of $262\frac{1}{2}$ lines per field. This simplifies circuitry and is referred to as *random interlace*. Many CCTV systems use random interlace when sync lock is lost. In general, random interlace does not degrade the picture.

Figure 7-2 shows a sync generator in block diagram. This may be included with the camera or it may be a separate unit. Most CCTV cameras include sync circuitry, although this may be bypassed and external sync, applied through a connector, can be provided. A master oscillator acts as the time standard. A stable crystal oscillator could also be used. In any case, the oscillator frequency is about 31.5 Hz for a 525 line rate. Other line rates could be used by changing the master oscillator frequency. The frequency in this case would be determined by multiplying the line rate by the field rate.

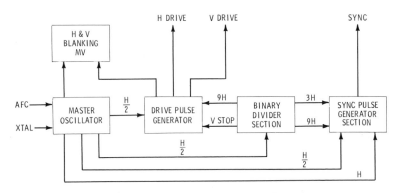

**Figure 7-2.** Sync Generator Block Diagram.

The master oscillator output is counted down IC J-K flip-flop multivibrators. The horizontal drive runs at one half of the oscillator frequency or 15,750 Hz. The countdown is then continued to 60 Hz. This 60 Hz is then compared to the 60 Hz line frequency, and an AFC circuit holds the master oscillator so that the counted-down output stage is in phase with line voltage. The camera-drive pulses trigger vidicon trace and blanks the beam during retrace. Both vertical and horizontal pulses are produced. These pulses have sharp rise times.

## THE COMPOSITE VIDEO SIGNAL

The camera must now produce a composite signal consisting of the video portion together with sync. The synthesis of one horizontal line is shown in Figure 7-3. At point 1 is the signal from the vidicon itself. At point 2 we add the blanking pulse. This pulse is from 10.48 to 11.3 microseconds long and is sufficiently wide to encompass the entire retrace cycle in the monitor, also allowing for delays due to circuitry or cable. At point 3 the blanking pulse is clipped approximately 0.1 volts below the level that will drive the monitor completely black. This extra 0.1 volt insures that the monitor will remain blanked despite overshoots, etc. due to circuitry. It is called *set-up* level. Finally,

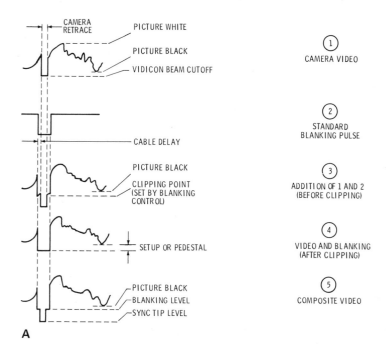

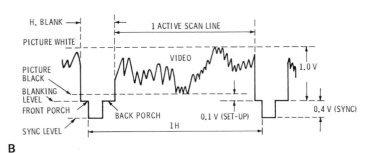

Figure 7-3. Once Composite Horizontal Scanning Line.
A, Synthesis. B, Composite.

at point 4 we add the sync pulse which will trigger the horizontal scan at the monitor. This results in the *front porch* and *back porch*.

The vertical sync components must also be added to obtain the entire composite signal. The vertical sync components are included in the output of the sync generator, which, as has been pointed out, may be a part of the camera circuitry, or may originate in a separate piece of equipment. To avoid repetition, we will include a discussion of the nature of vertical sync when discussing monitors.


## NIGHT VIEWING CAMERAS

Development of the silicon pickup tube makes possible viewing under very low lighting conditions. These tubes are used with the fastest possible lenses, those with apertures of f/1.8. For daytime viewing, the lens is stopped down and an attenuating filter is employed. Automatic iris adjustment is used almost universally. For conditions of total darkness or when reenforcement is desired, infra-red illumination is employed.

An infra-red illuminator is usually made from a tungsten lamp over which a filter is placed. The filter eliminates almost all visible light and only a dull red glow is observed. The assembly may be mounted on a pan and tilt drive together with the camera. An infra-red image does not appear the same as a visible light image. The image appears reversed and human features are difficult to recognize for the untrained eye. Nevertheless, infra-red is an excellent surveillance medium. Movement is easily recognized and certain objects appear in sharp contrast. The infrared source may be focused into a narrow beam for greatest distance. A source of three hundred watts, when employed with a fast lens and silicon diode pickup tube, will have a range of several hundred feet.

Another method of night viewing which is related to television but does not employ a conventional camera is known as light intensification. Using proper lenses this system is capable of viewing a man under starlight at distances up to one thousand feet. The method employs a light-intensifier tube which can amplify available light rays that form the image by a factor as high as 50,000. A fast (f1.4) long-photo-length lens focuses the image on the intensifier tube. Output of the intensifier tube may then go to another lens system where it can be observed by eye or placed on a conventional CCTV camera for remote viewing. An example of the performance of a light intensifier system is shown in Figure 7-4. This system, known as Star-tron, is produced by Smith & Wesson and offered to law enforcement and security agencies. The unit is battery powered for field use.

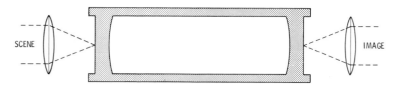

A

| Specifications | | 75 mm f/1.4 | 85 mm f/1.8 | 135 mm (2) f/1.6 | 135 mm f/1.8 | 170 mm (2) f/1.5 | 210 mm f/2.8 | 300 mm f/2.8 | 300 mm f/4.0 | 500 mm f/4.0 |
|---|---|---|---|---|---|---|---|---|---|---|
| **Moonlight (10-2 ft. candle)** | | | | | | | | | | |
| Man | — feet | 1380 | 1575 | 2280 | 2052 | 2625 | 2770 | 3870 | 3120 | 4020 |
| | — meters | 420 | 480 | 695 | 625 | 800 | 845 | 1180 | 950 | 1225 |
| Small Car | — feet | 1740 | 1968 | 2850 | 2560 | 3280 | 3450 | 4840 | 3900 | 5020 |
| (Jeep) | — meters | 530 | 600 | 870 | 780 | 1000 | 1050 | 1475 | 1190 | 1530 |
| Truck | — feet | 2425 | 2755 | 3990 | 3591 | 4592 | 4840 | 6760 | 5460 | 7020 |
| (Tank) | — meters | 740 | 840 | 1215 | 1095 | 1400 | 1475 | 2060 | 1665 | 2140 |
| **Starlight (10-4 ft. candle)** | | | | | | | | | | |
| Man | — feet | 1040 | 1180 | 1710 | 1560 | 1968 | 1950 | 2740 | 2030 | 2540 |
| | — meters | 315 | 360 | 520 | 475 | 600 | 595 | 835 | 620 | 775 |
| Small Car | — feet | 1300 | 1470 | 2130 | 1940 | 2460 | 2440 | 3430 | 2530 | 3182 |
| (Jeep) | — meters | 395 | 450 | 650 | 590 | 750 | 745 | 1045 | 770 | 970 |
| Truck | — feet | 1820 | 2070 | 2990 | 2730 | 3510 | 3410 | 4810 | 3545 | 4460 |
| (Tank) | — meters | 555 | 630 | 910 | 740 | 1070 | 1040 | 1465 | 1080 | 1360 |
| Absolute Field of View (19 mm Photocathode Diameter) | | 14.4° | 12.1° | 8.0° | 8.0° | 6.4° | 5.2° | 3.6° | 3.6° | 2.2° |

These figures were determined with an object having a contrast ratio of 30%. (A contrast ratio of 100% would be a white object against a black background. 0% contrast would be black against black.) An increase in target contrast from the base 30% will extend the range capabilities of the above lenses.

B

Figure 7-4. Performance of Light Intensifier. A, Tube. B, Performance of Smith & Wesson.

## MONITORS

A monitor accepts the normal TV signal and displays the image. Monitor circuitry is much the same as in an ordinary household TV receiver, but lacks the means of receiving VHF or UHF signals and detecting modulation components. The picture quality requirements of monitors vary. Monitors for use in color broadcast facilities must meet higher standards than monitors used with a lower priced monochromatic surveillance camera. Color monitors are the most complex and expensive. Most monitors used for surveillance use standard 525 line inputs and are built to accept external sync.

Input connectors are for either standard UHF or BNC connectors, and video input is provided with a 75-ohm shunt that may be switched on or off. Two or more monitors may be fed from the same video line but only one monitor (the last on the line) should then be set for 75-ohm input. The same applies to external sync.

Monitors for CCTV employ 5-, 9-, 14-, 17-, or 23-inch screens. The smaller sizes are most popular, since they are usually mounted in a console and the operator's eyes are fairly close. The four controls for vertical and horizontal size and hold are usually recessed or covered since adjustment is seldom needed. Focus controls may also be accessible from the front, but are also usually covered. Some monitors are actually TV receivers and contain tuners and demodulators which may be bypassed for CCTV. TV tuners are available that produce a standard composite video signal for feeding CCTV monitors. There are no audio stages in the ordinary monitor.

A simplified monitor block diagram is shown in Figure 7-5. The input has two connectors in parallel. When more than one monitor is fed from the same cable, one connector is used for a continuation of the cable. The last monitor in the series will have the termination switch in the 75-ohm position, thus terminating the line.

The incoming signal is first amplified, then the process is begun to separate the video from the sync signals. The sync stripper rejects the positive video portion and acts only on the negative sync. The resultant output is a composite sync signal. When external sync is used, a switch disconnects the incoming composite video from this circuit and replaces it with the incoming external sync.

The monitor sync separator circuit must first separate the sync from the composite video. The video portion is removed by a clipper circuit which passes only the negative portion below video

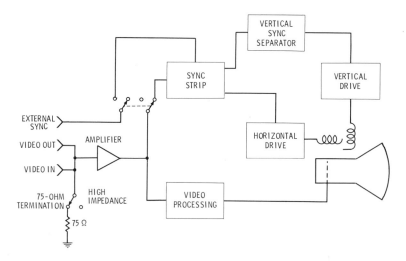

**Figure 7-5.** Simplified Monitor Block Diagram.

black level. The next function is to segregate horizontal from vertical sync. This is done by frequency-selective circuits.

Horizontal information is obtained by applying the composite sync signal to a differentiator circuit (Figure 7-6A). The time constant of the differentiator is selected to produce a train of pulses (Figure 7-6A) which are used to synchronize the horizontal sweep. Note that in both examples only the positive pulses of the differentiated waveform are used.

The vertical trigger pulse is obtained from an integrator circuit (see Figure 7-6B). The integrator circuit does not react to the short duration horizontal sync pulses. The integrator capacitors charge during each pulse and then discharge during the interval between pulses. The average charge of the capacitors is dependent upon the time constant selected for the RC combination.

Note that the integrator output waveform produced by the application of horizontal sync pulses during period "A" of the first field is shifted 0.5H in time from the waveform shown in period "A" of the second field. If the vertical sync pulses were applied at the beginning of period "B" in each field, the resulting vertical trigger pulse would be different on alternate fields due to the 0.5H shift in the waveform of the integrator at the end of period "A."

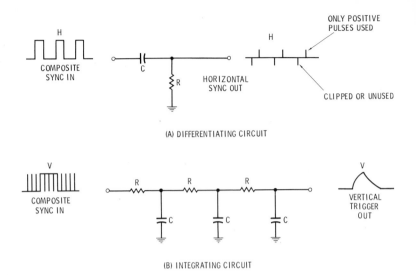

**Figure 7-6.** Circuits Used to Derive Horizontal and Vertical Sync.

With reference to Figure 7-7, note that application of an equalizing interval during period "B" provides a similar waveform for each field without a significant change in the average charge, since the duty cycles of the equalizing pulses and horizontal sync pulses are the same. The condition of the integrator in regard to the average charge and waveform, therefore, is nearly identical for the start of the vertical trigger at the beginning of period "C" for each field.

In addition to the latter function, the equalizing pulses serve another purpose. During the vertical interval, without horizontal sync the monitor would lose horizontal synchronization. Alternate equalizing pulses occur at a rate equivalent to that of horizontal sync pulses, however, thus preserving sync. Also, the leading edge of each alternate equalizing pulse occurs at a position in time at which the leading edge of a horizontal pulse would normally occur. The necessary horizontal synchronization is obtained from the horizontal sync pulses when applied to a differentiator circuit as shown in Figure 7-6A.

Horizontal and vertical sync initiates the respective sawtooth sweep. Because of tube geometry, the sweeps should not be linear. To compensate for the deflection angle of the kinescope

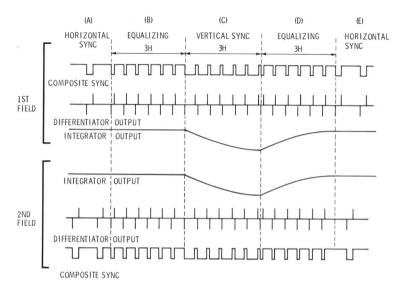

**Figure 7-7.**   Processing the Equalizing Pulse.

and the fact that the screen is not a uniformly radial distance from the beam deflection point, sweeps are compensated by special circuitry. Two kinds of horizontal linearity correction are usually employed. To compensate for circuit losses, the raster should be compressed on the left and expanded on the right. A parabolic tendency is added to compensate for the fact that the screen is flatter than a sphere centered at the point of deflection.

Each monitor manufacturer may employ different circuitry to accomplish the various functions. In general, the ability of a monitor to produce a good picture at lower signal levels is the most significant criteria. Resolution is usually best at the center of the screen, but there should not be excessive drop-off at the edges. When fed from a good camera, the gamma should remain good.

## VIDEO TAPE RECORDING

Storage of a video signal, especially temporary storage, is a great advantage in certain surveillance situations. The advantage of tape over conventional motion pictures is that a taped scene is available for instant replay. Also, tape may be erased and used again. Different camera outputs may also be placed in succession on the same tape. Video tape recording required a large development commitment before it was perfected. The reason for this is the wide spectrum occupied by the video signal. Video tape recorders (VTR) offered now are all based upon the same principle, but a variety of systems are used by the different manufacturers.

Magnetic tape consists of fine particles of an iron that can be easily magnetized and retain a polarity. The particles are deposited on the tape by a number of processes developed by various manufacturers. A small area of tape may be magnetized to a degree relative to the adjacent area, and the resulting differences become the recorded signal.

Iron will become magnetized when a conductor carrying current is adjacent to it. This is the recording process. When the magnetic lines of the magnet move so as to cut through a conductor, a voltage is induced. This is the playback process. Thus, recording is a function of *current* and playback is a function of *tape velocity*. Also, in video recording, impulses change at a rate up to 4 MHz, and this requires a very fast recording speed to get the signal on the tape. Tape must move past the recording head at speeds of hundreds of inches per second to reproduce the highest components of a signal. There is a problem, however. When a signal is played back, it is the *rate of change* of magnetic lines of force that causes a signal voltage to be induced. There are very low frequency (60 Hz) components in a television signal, and these would be induced with much less magnitude than the high frequency components. In fact, they would be lost in inherent circuit noise.

A recording system, therefore, cannot be made that will pick up a very wide range of frequencies. For a given recording system, doubling the frequency increases the induced output 6 Db, and halving the frequency decreases output by 6 Db. Although audio recording is quite successful, this tells us that the wide frequency range of TV signals simply cannot be recorded in the same manner. This limitation is solved, however, by recording a high–frequency carrier signal, and frequency-modulating the carrier. The frequency excursions of the carrier are made to be within the recording capability of the system. Although the frequency response of the recording and playback process is not linear, this is overcome merely by limiting or clipping the impulses to a uniform level.

## Helical Systems

The recording head of any tape recorder is what is known as a magnetic circuit. Magnetic lines travel around the core as shown on Figure 7-8A. A gap in the magnetic path, as shown, can induce magnetism into the adjoining tape by allowing the tape itself to complete the magnetic circuit. The gap in a VTR must be made as small as possible in order to crowd the recorded impulses as close together as possible, and may be formed of a nonmagnetic material rather than air. In order to record a single cycle of a 1 MHz wave, one microsecond must be used to record the entire cycle. This means that the tape must move through the influence of the gap in that time.

Another way to say this is to relate the speed of the tape past the head with the recorded frequency. Using inches per second for tape speed:

$$\text{Length of tape portion to record one cycle} = \frac{\text{Inches per second}}{\text{frequency}}$$

The portion of tape used to record one cycle is related to the gap width. The smaller the gap, the less the length—down to certain practical limits. Once this limit has been reached, only by increasing tape speed can the frequency be increased.

Tape speed is relative. That is, either the tape can move or both tape and head may move in relationship to each other. In a video recorder, the recording head is designed to revolve as the tape moves past. In most recorders, one full field or one full frame is recorded in one revolution of the head. There is still another thing left that can increase writing speed, however. This is called *helical scanning*. With reference to Figure 7-8B, note the angular relationship of scan and tape travel. The relative speed is now the vector sum of the two velocities, which is considerably greater than the relative peripheral velocity of the scanning drum and linear velocity of the tape.

Summarizing, then, video tape recording, with its wide spectrum, is possible because;

1. the video signal is frequency modulated on a carrier.
2. the recording head itself is in motion relative to the tape.

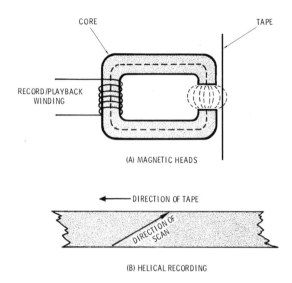

**Figure 7-8.** Video Tape Recording.

In addition, helical scanning allows higher writing speeds with simple mechanisms.

Several manufacturers in the United States, Europe, and Japan now offer VTR's, each with their own system. Recently, VTRs have been introduced for use in surveillance. They are characterized by a means of recording individual frames at specified intervals or alternate frames from different cameras (see Figure 7-9). This is called *time lapse* recording. This is done by using a common sync and switching only during frame or field interval. Character generators can be introduced to identify the frames. This will be discussed under "Switching and Special Effects."

An example of a VTR designed for surveillance is the RCA 3350. This unit can be set to record individual frames at intervals so that a two-hour tape will last up to 300 hours. When the tape is played back, we see a succession of scenes taken at intervals of minutes, depending upon the setting. This is time lapse recording. *Real time* recording is the scene in ordinary motion. This model is designed so that an external alarm will cause the recording mode to go from time lapse to real time. One hour video cassettes are used for ease of loading and storage. An audio signal may also be recorded or dubbed in later.

**Figure 7-9.** Time-Lapse Video Cassettes Recorder (VCR) with Built in Date/Time Generator. (Courtesy of RCA.)

# 8

# *Lenses*

The television industry traditionally speaks of the camera optics as a *lens*, although it is actually made up of several lenses or elements which are carefully mounted in relationship to each other. When referred to here the term "lens" refers to the entire assembly, sometimes also called a *compound lens.*

Cameras may be ordered with or without a lens. When ordered without a lens, the CCTV installer makes an individual choice of a lens that depends upon his situation. In order to make cameras and lenses interchangeable the industry has standardized on a thread size and diameter for the rear portion of the lens assembly which screws into the camera. The camera mounting is referred to as a "C" mount. The diameter selected is one inch, with a thread pitch of 32 per inch. The standard usually applies to foreign as well as domestic made lenses. Lenses made for photographic purposes also comply with this standard.

The lens operates by projecting a scene on the sensitive portion of the camera tube or other light sensitive device. A lens may be designed to "see" different amounts of the area placed before a camera. One way to classify a lens is by the angle of the scene a lens subtends. A lens may be *telephoto* or *wide angle.*

## OPTICAL CONSIDERATIONS

Grinding individual lens elements is both an art and science. German lens manufacturers were the first to produce large quantities of fine lenses. Techniques soon followed in the United States and other parts of Europe. Finally, Japan has emerged as a mass producer of precision lenses.

A compound lens such as used in CCTV and photography consists of several elements, each with differing optical characteristics. The lens elements in combination are made to correct for chromatic aberration, spherical astigmatism, and curvature of field. A lens production process will yield only a few elements that meet the highest standard. Certain types of lens elements are more difficult to grind. The manufacturer selects elements and produces a line of lenses that utilize elements of various gradation, carefully chosen for optimum performance in a given function.

Fortunately, the nature of television is such that a lens intended for fine photography need not be used. Most aberrations that affect a fine photographic lens are tolerable to a greater degree in television with the exception of distortion (curvature of field). Distortion is generally grouped under two classifications, as shown in Figure 8-1. When evaluating a lens in terms of resolution, the effect of the lens on entire system resolution must be considered. Figure 8-2 shows the aggregate system resolution versus lens resolution. From the graph, it appears that when lens

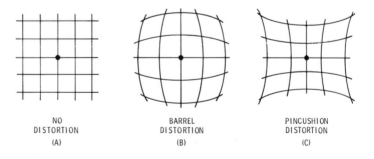

NO DISTORTION (A)          BARREL DISTORTION (B)          PINCUSHION DISTORTION (C)

Figure 8-1.  Types of Distortion.

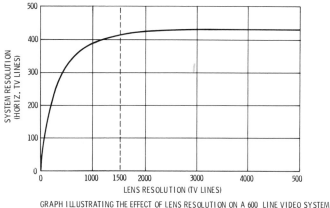

GRAPH ILLUSTRATING THE EFFECT OF LENS RESOLUTION ON A 600 LINE VIDEO SYSTEM

**Figure 8-2.** System Resolution Versus Lens Resolution.

resolution exceeds about 1500 lines, improving lens resolution will result in little improvement in system performance. Most lenses will have resolutions better than 1000 TV lines. It happens that the lens will exhibit better resolution and contrast toward the center of the image. The relationship of resolution versus contrast is known as *modulation transfer function* (MTF), and the MTF will be better at the image center than at the periphery.

## FOCAL LENGTH

Cameras for CCTV are standardized for two sizes of pick up tubes. That is, the focused image falls upon a sensitized area of two standardized sizes. These we know as *formats*. In both cases, the area has the traditional 4:3 aspect ratio. The smaller format is known as *2/3 inch* or *2/3 format*. The larger is simply called *1 inch*. Neither of these dimensions are exact. The dimensions are as shown in Figure 8-3. It is important to distinguish between the diagonal dimension and the length and width dimensions. The scene coverage for a given lens has to do with the relationship of its focal length with format size.

Focal length determines the angle of coverage of a lens. It is the distance from the theoretical optical center of a lens to the

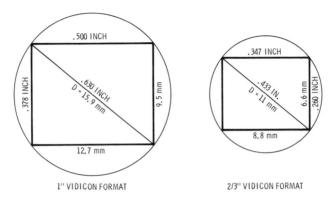

Figure 8-3.  Standard Image Formats.

focused image on the light sensitive surface of the camera pickup tube. The amount of area offered by the camera tube may be in one inch or 2/3 inch format, but this is independent of focal length. An in-focus image will be the same distance from the optical center of the lens regardless of the amount of image space actually utilized, as shown in Figure 8-4. The smaller image area of a 2/3 inch camera merely means that there will be less scene coverage.

The amount of scene coverage for a given camera location is important in video surveillance. We note from Figure 8-4 that scene coverage is a simple geometric proportion dependent upon focal length and whether we use a one inch or 2/3 inch camera. Because we are dealing in simple proportional triangles, we can work out the simple proportional relationship:

$$\frac{\text{Format width}}{\text{Focal length}} = \frac{\text{Scene width}}{\text{Distance from Camera}}$$

or we could also say:

$$\frac{\text{Format height}}{\text{Focal length}} = \frac{\text{Scene height}}{\text{Distance from camera}}$$

In dealing with these relationships it must be remembered that all quantities must be in the same units of measure. But focal lengths are given in millimeters, format in inches, and camera

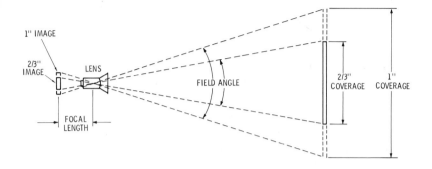

**Figure 8-4.** Focal Length Geometry.

distances and scenes are most conveniently measured in feet. The reader in possession of one of the new pocket calculators will have little trouble with this, however. Use feet for camera distance and scene width, then:

For one inch cameras:

$$\text{Distance from camera–scene width (height)} \times \frac{12.7}{\text{focal length (mm.)}}$$

For 2/3 inch cameras:

$$\text{Distance from camera-scene width (height)} \times \frac{8.8}{\text{focal length (mm.)}}$$

The figure of 12.7 and 8.8 came from Figure 8-3.

It can be seen that a given lens will work on either the 2/3 or the one inch format. The only difference is that a 2/3 inch format will result in decreased coverage. Nevertheless, manufacturers will occasionally state that a given lens is for use on one or the other. This is because many CCTV lenses were originally designed for photographic use, and redesigned in part for CCTV, and a smaller format usually results in simpler design for given quality image. In general, the use of a given lens on the smaller format will give a better image. This is because image quality increases toward the center of the format.

An important consideration in locating a surveillance camera is the amount of lens coverage necessary to recognize an individual. The camera must either be placed close enough that the subject can be recognized, or a longer focal length lens employed. Although short focal length will cover a greater area, the subject under view must be closer to the camera to recognize features. A general rule for recognizing human features is to assume a camera distance of one foot for every millimeter of focal length. The only sure check, however, is to make an actual observation or simulate a system. This is because of the interplay of resolution and lighting for a given situation.

## LENS SPEED

This is the ability of the lens to collect light from the scene. The word "speed" is a hangover from photographic terminology, in that more light meant faster shutter speeds. A video security system is governed by its ability to reproduce an image at a given distance under given light conditions. The image must be sufficiently recognizable to satisfy surveillance requirements. An image may be close or far from the camera, and a wide range of lighting conditions may prevail in the area under observation.

The CCTV camera itself usually operates over a comparatively narrow range of light conditions, and is limited by the characteristics of the vidicon or other pickup source. It is therefore necessary to employ a diaphram to adjust the lens so that the camera may function over a wider light range. This is done by starting with a lens and camera combination that will provide a satisfactory image under the least light expected for a given situation. As light increases, the lens is *stopped down* so that the camera receives roughly the same amount of light. In general, therefore, the f-stop and light transmission of a lens is the most important characteristic that governs performance of a video system that is required to operate over a range of lighting conditions.

Lens speed, or f-stop, is the ability of the lens to collect light from the scene. In a theoretically perfect simple lens made up of a single element of glass, f-stop is designated as

$$f = \frac{\text{focal length}}{\text{lens diameter}} \quad \text{(aperture)}$$

This quantity is known as *f-stop*. Because a CCTV lens is actually a compound lens, the aperture does not equal the diameter of the front element of glass of the system. We therefore have an *effective* aperture (also sometimes called an *apparent* aperture) which is the opening that is equivalent to the largest beam of parallel light transmitted by the lens. The effective aperture lies somewhere within the lens structure depending upon design.

The diaphram is usually an iris arrangement of opaque thin metal that closes and opens to allow regulation of the light, hence, changing the aperture of a given lens. In the system of f-numbers, each narrower diaphram setting (or aperture) reduces the amount of light entering by one half of the previous setting. Conversely, each wider diaphram setting doubles the amount from the previous setting. Figure 8-5 shows how the diaphram opening changes to reduce the amount of light gathering area. If the diaphram is a circle, the amount of light is a function of the area of this circle. Then the area of each circle is reduced by a factor of one-half as the diaphram is closed.

The f-stop is the ability of the lens to gather light. The amount of light is inversely proportional to the square of the f-stop. We could write this as

$$X = Y \frac{f_1^2}{f_2^2}$$

where X is the amount of light transmitted with the diaphram set at $f_2$ and Y is the amount of light transmitted with an f stop value of $f_1$. Light values may be in lumens or any unit so long as the same units are used, or the relative calibrations of a light meter or simple photo cell may be used.

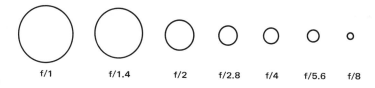

| f/1 | f/1.4 | f/2 | f/2.8 | f/4 | f/5.6 | f/8 |

**Figure 8-5.** Relative Iris Diameters. This illustration shows the relative iris diameters for a theoretical lens in which the aperture is equal to its focal length. The area of each circle is one-half that of the larger.

Notice that we are speaking in relative values for a given lens. This does not mean that two different lenses will transmit the same amount of light when the diaphram is set to say, f/2. The reason for this is that many factors within the lens affect its ability to transmit light which has been gathered from the scene. Light is lost in the imaging process due to characteristics of the optical glass, coatings, mechanical design, etc. For this reason, the percent of transmittance is used as a factor in lens performance.

Transmittance is determined from actual measurement of a given lens. It is a comparative measure of luminance, or the ratio of transmitted light to incident light. This ratio will always be less than one, since there will always be light loss. In general, a good rule of thumb is that a fixed focus lens will lose about one-half f-stop from light transmission loss, and a zoom lens will lose one f-stop.

The face plate sensitivity of the vidicon tube will govern the limits at which a given lens performs under a range of scene lighting. Most CCTV cameras are rated in face plate illumination. This is the amount of light required on the vidicon faceplate itself to gain full performance. If a scene is brighter, a smaller aperture (higher f-stop) setting is used, and vice versa. The amount of faceplate illumination for a given scene can be calculated by the equation:

$$C = \frac{BT}{4f^2}$$

where

C = faceplate illumination in foot-candles
B = scene brightness in foot lamberts
T = percent transmittance of lens
f = aperture setting in f-numbers

For example, if the scene brightness is 40 foot lamberts, with an f-stop of 1.4 and a lens which transmitts 80 percent of the light, the faceplate illumination will be:

$$C = \frac{40 \times .8}{4 \times (1.4)^2} = 4 \text{ foot-candles}$$

The f-stop or aperture setting of a lens has another important consideration: it determines the depth of field. The depth of field is the range of distance at which an object will appear in focus. The narrower the aperture, the greater the depth of field, although greater scene brightness will be required.

An iris assembly may be used to increase the depth of field. When more light is available a lens may be *stopped down*, that is, the iris may be closed to allow less light. This has the effect of broadening the depth of field, causing sharper images throughout the desired viewing range. A camera operates over a fairly narrow range of light values, hence, a camera capable of operating at low light levels, when operating with normal light and with iris narrowed to high f-stops, will give a sharp image over a wider depth of field.

For a given lens assembly and iris setting, if all objects from, say, 30 feet to 50 feet remain in satisfactory focus, the depth of field may be said to be 20 feet. The figure is arbitrary, since it involves human judgement and the design of the entire video system.

## LIGHT ATTENUATORS

Modern cameras can be made to operate at extremely low light levels even when no means of image intensification are employed. These very sensitive cameras cannot operate in bright light, however, without some means of attenuation even when the iris is stopped down as far as practical. By employing filter attenuation, however, a system can be made to operate under conditions from bright sunlight to partial darkness.

One method consists of a conventional iris assembly to which a neutral density filter has been added. A motor or solenoid driven dual position filter arm carries either the filter or a clear optical flat into the optical path, depending upon lighting conditions. The filter is driven in or out of position automatically at either end of the normal iris range. The clear optical flat is placed into the light path when the filter is out, in order to compensate for path-length difference between air and glass, thus preserving focus. If the filter is out of position and greater light attenuation is required, the filter will automatically drive into position at the fully closed iris setting. Conversely, if the filter is in position and more light is required, the reverse occurs at the iris full open position. Thus, a dual iris range is achieved.

Iris range achieved by this method is approximately 150 to 1. Effective filter attenuation is approximately 50 to 1. Therefore, the practical combination of iris and filter is approximately 7,500 to 1.

The main advantage of this system is maximum light transmission with fully opened lens aperture. Disadvantages are slower response time to certain lighting changes and more complex mechanical assembly.

Another method consists of a filter with increasing density toward the center of the optical path. This is called a *spot filter* (see Figure 8-6). As the iris closes it subtends the denser filter area with increasing light attenuation. At larger iris openings the effect of the spot filter is almost negligible. Thus, the lens allows a sensitive camera to operate under conditions of partial darkness, yet permits operation at high light levels when the iris is narrowed to the attenuated zone. Operating light range of the lens is uninterrupted as the iris is driven from highest to lowest lens f-stop. This method can be made to attenuate to an f-stop number of about 360.

Advantages of the spot filter method are minimum response time with continuous iris range, increased attenuation, and simple construction with no moving parts. A disadvantage is a slight light loss from 1/4 to 1/2 f-stop at full-open aperture.

Figure 8-6.  A Motorized Intraspot™ Lens.

## ZOOM LENS

The zoom lens is a cleverly designed mechanical assembly of lens elements arranged in such a way that they can be moved to change the focal length from wide angle to telephoto while the image remains in focus.

A zoom lens is rated by the range of focal lengths over which the lens operates. For example, a zoom range from a 20 mm focal length to a 100 mm focal length would be 5x, a zoom range from 15mm to 150mm would be 10x, and so on.

In general, the greater the focal length range and lower the aperture rating, the more expensive the lens. As can be expected, there are design tradeoffs in a lens. A well built 4x zoom lens will usually have less distortion, and greater speed and resolution than a 10x lens made with the same care.

A focusing adjustment is provided on zoom lenses to allow for overall tolerance. Once focused, a zoom lens should remain in reasonable focus throughout zoom range. Zoom lenses are designed for the standard .690 distance from "C" mount banking surface to the image surface. This distance should be as exact as possible for the image to remain in focus throughout zoom.

A typical zoom lens may be rated as follows:

- Aperture (f/stop range)
- Range (4x, 10x, etc.)
- Focal Length Range, mm
- Field Angles (at both telephoto and wide-angle setting)
- Minimum focusing distance

Field angles are rated either on the diagonal or through the format. It is important to distinguish the difference.

## MOTORIZED LENSES

Cameras which are remotely located cannot be adjusted by an operator. For remote adjustment, miniature electric motors are mounted on the lens. The usual method is to fasten ring gears on each adjustment barrel and to drive the gears through pinions (see Figure 8-7). Slip clutches are usually employed to slip the

Figure 8-7.   A Motorized Zoom Lens.

motor when an element is driven against the mechanical stop. By
employing permanent magnet DC motors, a function can be
driven in the reverse direction merely by changing polarity of the
energizing current. These motors are easily driven by a transistor,
although any DC source is satisfactory.

   The design of a lens drive takes into consideration the torque
required, which is often not uniform as the barrel moves. The
ring gear is usually polyurethane to provide metal-to-metal
isolation between lens and motor. Running should be smooth
and silent as possible. The presence of RF noise may disturb the
camera image and should be eliminated by filtering or other
means. Motor speed is usually varied by reducing driving voltage.
To insure uniform speed by this method, a simple DC amplifier

can be used. Amplifier input current, usually by a potentiometer, is thus isolated from the motor as a function of amplifier current gain. The driving source, called a *control*, is located at the control center. The control is housed for a desk top or panel mounting with control buttons accessible to the operator. Separate control buttons are employed for focus, zoom, or iris adjustment. A motor is geared to each (see Figure 8-8).

A fixed focal length lens is occasionally employed with a motorized iris. The reason for this is that variable lighting conditions often require that the iris of a remote camera must be adjusted. In some cases a motorized focus is also provided. The most common motorized lens is the zoom lens because a zoom lens is most

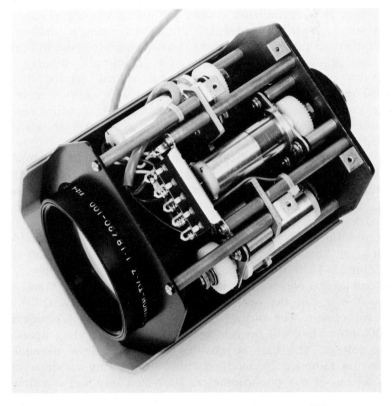

**Figure 8-8.** Three DC Motors Provide Adjustment of Zoom, Iris, and Focus.

often employed on a pan and tilt unit and requires adjustment for many different camera scenes.

A common method of driving a motorized lens is shown in Figure 8-9. Two amplifier sections are employed, one for each driving polarity, hence, direction. Two NPN transistors comprise the iris close, zoom-in, and focus-near, and two PNP transistors form the iris-open, zoom-out, and focus-far sections. Motor M1 drives the focus, M2 the iris, and M3 the zoom. Potentiometers R1 and R4 control the speed of the motors.

## AUTOMATIC IRIS CONTROL

Automatic iris adjustment to varying light conditions is a widely used feature in surveillance systems. By using a means of attenuation with a sensitive camera, the camera will operate unattended both night and day. This method is used on bridges and toll gates as well as in warehouses, vaults, etc. Electronic circuitry for auto-iris operation is fairly simple and a small printed circuit board may be incorporated in the lens itself. Another method is to contain the circuitry at the control point, in which case a switch may be used to convert from automatic to manual iris adjustment.

An auto-iris system utilizes the camera video output. It happens that the average value of a video signal is almost a direct function of the amount of light on the scene. The presence of sync pulses does not appreciably alter this. It is thus possible to use the video signal to adjust the iris by servo action. The video signal is converted to an average DC value end used to drive a DC amplifier which, in turn, energizes the miniature motor driving the iris.

Figure 8-10 is the schematic of a typical auto-iris drive. Q1-Q6 comprise a balanced push-pull DC amplifier. Q1, Q2, and Q3 drive the iris in the open direction and Q4, Q5, and Q6 drive the iris closed. The output is balanced to ground. Q7 provides a DC input that will be zero at some point depending upon its base voltage. The base is provided with a DC level depending upon the value of the rectified video supplied by the diode and the setting of the potentiometer. The video amplifier is a simple circuit consisting of one or two transistors and has a high input impedance that will not load the camera output.

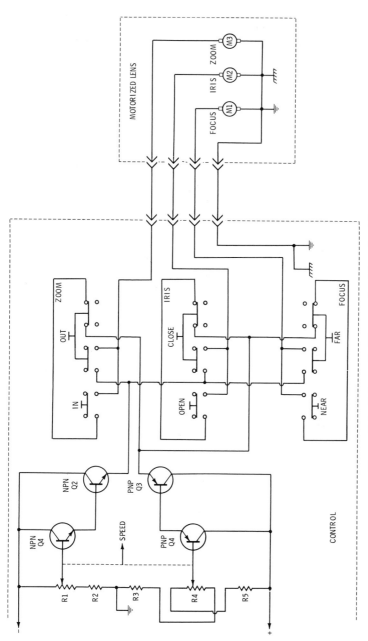

**Figure 8-9. Schematic Diagram for Motorized Lens.**

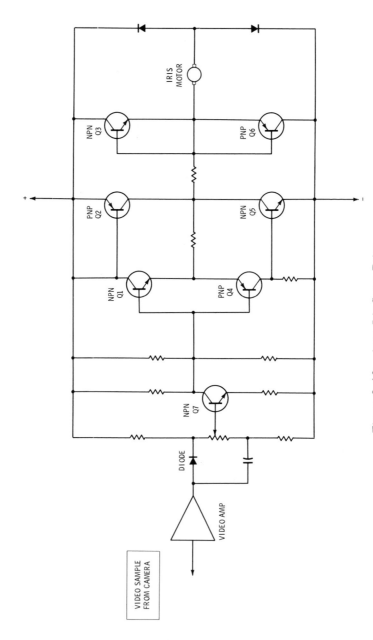

**Figure 8-10.** Auto-Iris Servo Drive.

When the camera signal goes below a certain level, Q7 will drive the amplifier output positive, causing the motor to open the iris. Conversely, when video becomes higher than a certain value, Q7 drives the amplifier negative, driving the motor in the opposite direction and closing the iris. The potentiometer setting determines the operating light level.

# 9

# *Video Switching*

A major feature of video surveillance is the advantage of switching outputs of several cameras from one or more control points. Any combination of cameras may be employed. A deterrent arrangement of cameras may be combined or switched to an apprehension system. Unused or out of service cameras may be bypassed, or certain cameras may be automatically monitored or switched by a variety of intrusion alarms ranging from video motion detection to conventional photocell detectors, sensors, and so on. Video switching may be in the form of a simple unit, or it may be a sophisticated solid state device with complex features.

CCTV switching requires that the effective video spectrum of a signal (about 4 Mhz.) be switched without causing "ghosts," high frequency loss, low frequency phase shift, or AC hum bars. In simple installations, this can usually be accomplished by simple passive switchers, providing cable runs are relatively short and proper terminations are made. In situations where video must be monitored and switched from several locations, it will be necessary to employ some form of signal processing to guarantee good picture definition and performance with all switching combinations.

The series of diagrams shown in Figure 9-1(A-F) show how camera scenes can be switched to a monitor.

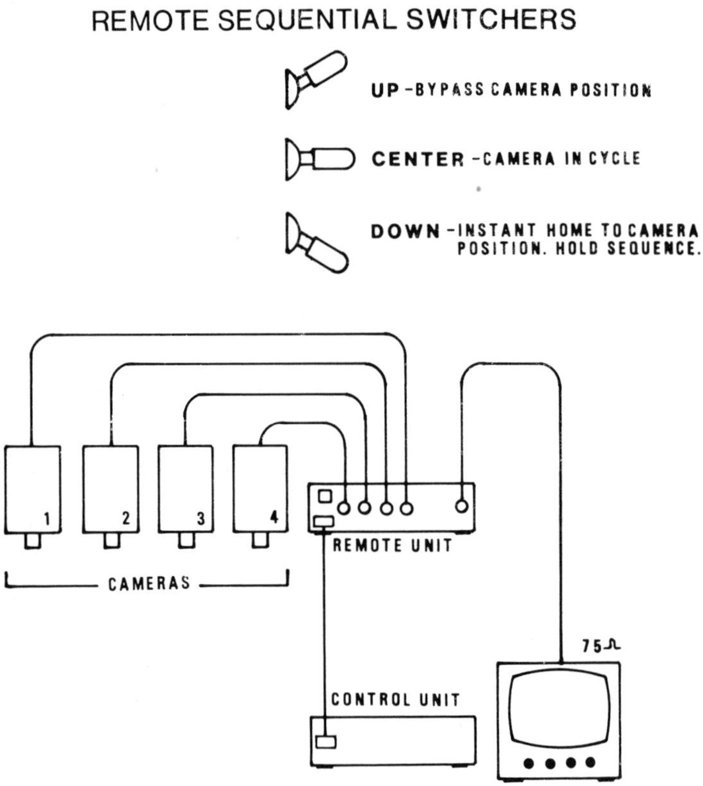
REMOTE SEQUENTIAL SWITCHERS
UP -BYPASS CAMERA POSITION
CENTER -CAMERA IN CYCLE
DOWN -INSTANT HOME TO CAMERA POSITION. HOLD SEQUENCE.
1
2
3
4
CAMERAS
REMOTE UNIT
75Ω
CONTROL UNIT

Figure 9-1A. Types of Video Switching. Remote Sequential Switchers.

## AUTO HOMING SEQUENTIAL SWITCHERS

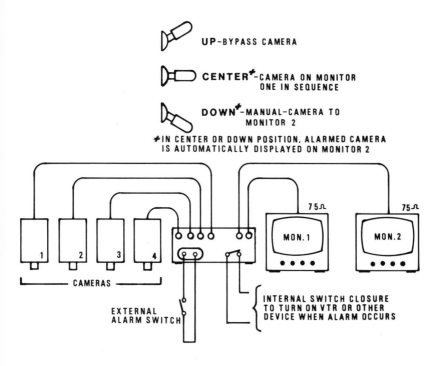

UP-BYPASS CAMERA

CENTER*-CAMERA ON MONITOR
ONE IN SEQUENCE

DOWN*-MANUAL-CAMERA TO
MONITOR 2

*IN CENTER OR DOWN POSITION, ALARMED CAMERA
IS AUTOMATICALLY DISPLAYED ON MONITOR 2

75Ω        75Ω

MON.1        MON.2

CAMERAS

INTERNAL SWITCH CLOSURE
TO TURN ON VTR OR OTHER
DEVICE WHEN ALARM OCCURS

EXTERNAL
ALARM SWITCH

**Figure 9-1B.** Types of Video Switching. Auto Homing
Sequential Switchers.

# TTL
# LOOPING INPUT BRIDGING SEQUENTIAL SWITCHERS

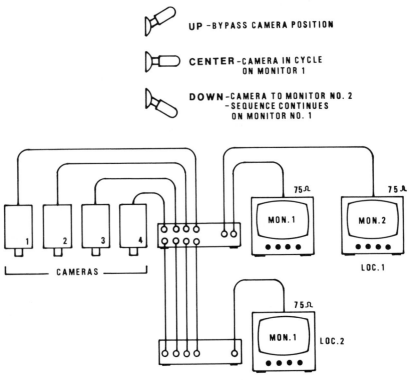

**UP** –BYPASS CAMERA POSITION

**CENTER**-CAMERA IN CYCLE
ON MONITOR 1

**DOWN**-CAMERA TO MONITOR NO. 2
–SEQUENCE CONTINUES
ON MONITOR NO. 1

**Figure 9-1C.** Types of Video Switching. TTL Looping Input Bridging Sequential Switchers.

## BRIDGING SEQUENTIAL SWITCHERS

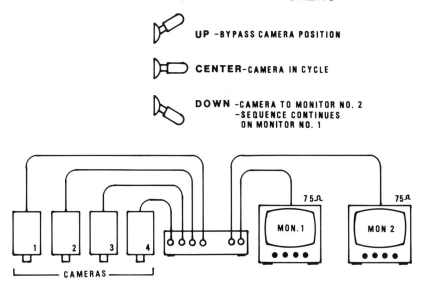

Figure 9-1D. Types of Video Switching. Bridging Sequential Switchers.

## TTL
## LOOPING INPUT HOMING SEQUENTIAL SWITCHERS

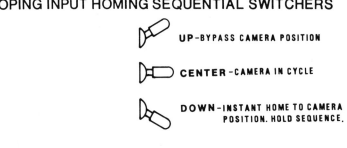

**UP-BYPASS CAMERA POSITION**

**CENTER-CAMERA IN CYCLE**

**DOWN-INSTANT HOME TO CAMERA POSITION. HOLD SEQUENCE.**

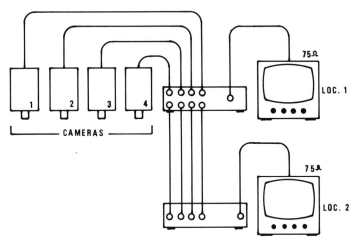

**Figure 9-1E.** Types of Video Switching. TTL Looping Input Homing Sequential Switchers.

# HOMING SEQUENTIAL SWITCHER

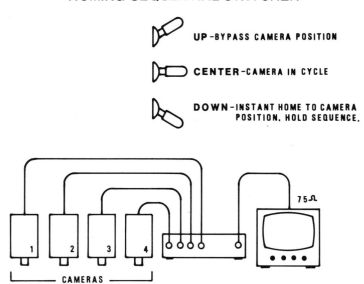

**UP-BYPASS CAMERA POSITION**

**CENTER-CAMERA IN CYCLE**

**DOWN-INSTANT HOME TO CAMERA POSITION. HOLD SEQUENCE.**

**Figure 9-1F.** Types of Video Switching. Homing Sequential Switcher.

## SWITCHING PRINCIPLES

The basic form of any video switching arrangement is shown in Figure 9-2. Here, four single-pole double-throw switches are formed together in a single unit. Although four are shown, any number may be used. The important feature is that the switches are interconnected mechanically so that only one switch may be in position B at a time. The function may be accomplished mechanically or it may be by solid state means, but the principle is the same. The arrangement shown may be packaged in distinct ways to provide the desired function.

Certain terminology has come into use to describe the various switching combinations and functions. A *passive* switcher is one which switches the signal without processing it in any way, and the switching is carried out without disturbing proper cable

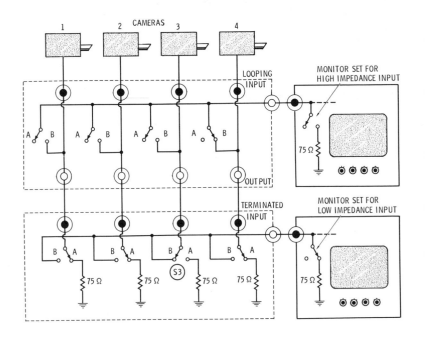

**Figure 9-2.** Basic Video Switching Arrangements.

termination. It is also usually understood that simple passive switchers employ common mechanical contacts, often using switches originally designed for low voltage DC applications. It will be recalled that the majority of surveillance systems employ unbalanced coaxial lines with a characteristic impedance of 75 ohms. The cable, which has a characteristic impedance of 75 ohms, must be shunted at its terminal end with 75 ohms under all situations. Otherwise ghost or spectrum loss will result.

A monitor, switching unit, or other device is said to be *looped* into a circuit when the circuit passes through the device without disturbing the respective terminating impedance. Thus *looping* is a simple "T" connection (providing that the device in the loop is set so that it offers negligible impedance). This is the purpose of supplying a switch on most monitor chassis so that a 75 ohm resistor may be switched on or off across the input.

The term *bridging* is used simply to indicate that a device is capable of being switched across circuits. In all cases, the process of looping or bridging must maintain the termination of a circuit. This can be accomplished as shown in Figure 9-2. Notice that here again Switch No. 3 (S3) is in Position B, hence this circuit is in view on the monitor. All other circuits will be across the respective terminating resistor. The switch on the monitor chassis is set for a 75 ohm impedance. Due to internal interlocking of switches, all circuits remain terminated. The question may be asked, "Why bother to terminate a circuit which is not being viewed?" In simple cases, there is no reason. Termination is habitually supplied, however, so that additional switches for bridging to another monitor may be looped into the circuit if so desired.

The principle of looping and bridging is so simple that it may appear complicated from over explanation. As surveillance systems become more complex, employing more and more switchers, it is handy to diagram a switching system as shown in Figure 9-3. This way of simplifying switching arrangements has been adapted from space center and broadcast TV systems. Signal flow is usually shown vertically and bridging horizontally. Thus the looping (vertical) path is at right angles to the bridging path.

So far nothing has been said of the necessity of signal processing. It is implied that all switchers have a passive function. There is, of course, a limit to the number of passive switchers that may

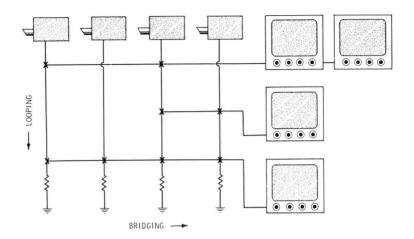

**Figure 9-3.** Looping and Bridging Switching Diagrams.

be added before some processing is necessary. Long cable runs between switchers at different locations may require isolation, high frequency boosting, or separate termination. This can be accomplished without losing flexibility or ease of future system modification by using video signal equipment that is separately housed and introduced into the circuit by standard connectors. The physical arrangement may be racked, set into a console, or used with individual desk-top mounting. Rearranging or changing a security guard station, for example, may thus be accomplished quickly even by untrained personnel.

## SWITCHING TECHNIQUES

When servicing video switching equipment, it is a good idea for the technician to understand the method and locate the point at which actual switching occurs. The term *solid state* switching as used in advertising literature means that a diode or transistor is used to cause a direct current or video signal to be turned on or off or to change path.

One method is to use a diode. The effect of varying the voltage across a resistor is shown on the graph of Figure 9-4A. This

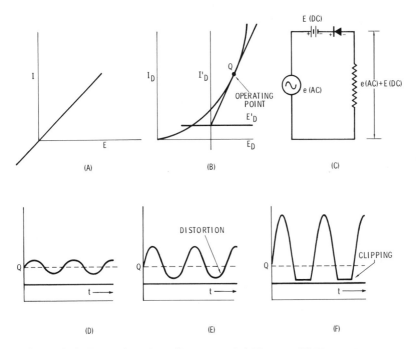

**Figure 9-4.** Diode Switching Principles. (A) Voltage (E) Versus Current (1) Curve of Resistor (Ohm's Law). (B) Voltage Versus Current Curve of Diode. For a Small Portion of the Curve, the $E_{n'} - I_{n^-}$ May be considered linear. (C) Simplified Circuit to Produce DC Diode Bias, E (DC), and Carry an AC or Video Signal. The AC is Switched Off When E (DC) Changes Polarity. (D) Diode Current with Small AC of Video Signal superimposed on DC. As with Circuit Shown in (C), DC Level is Correct for Switching. (E) When the AC for Video Signal Increases, Distortion Begins to Appear. (F) Clipping Begins with Larger Video or AC Levels.

follows Ohm's Law and results in a straight line. Across a diode, the result is a curved line which becomes zero when the polarity changes. If we choose a point such as Q on Figure 9-4B, a very small portion of the curve may be considered a straight line. Figure 9-4C shows a simple circuit carrying both AC and DC. So long as the AC component is kept small as in Figure 9-4D, an undistorted AC component appears in the circuit. If AC is increased, however, distortion begins to take place as shown in Figure 9–4E. This is because the voltage–current relationship of the diode is not linear except for small portions of the curve. For still larger AC values, we note that clipping takes place (see Figure 9-4F). This is because current cannot flow in the reverse direction. If, now, the DC bias is reversed, the AC component ceases to flow entirely. This is the way in which an AC or video signal may be switched off.

Silicon diodes are preferred to switching and the DC switching bias may be supplied in the form of a square wave. Switching speeds may be made very fast, and video may be switched between frames or at any point throughout the sync intervals. Virtually no distortion is produced on one volt video signals when proper diodes are used. The diode method of switching is therefore a preferred method.

Transistors may also be used as switches. The technique is generally the same as with diodes in that turnoff occurs when the transistor base goes below cutoff. The base is driven below cutoff by a switching signal. The signal to be switched off rides on the switching, or bias, signal. Field–effect transistors (FET's) also may be used as switches.

A field–effect transistor is made from a sheet of N–type silicon as shown in Figure 9–5A. Current flows in either direction through

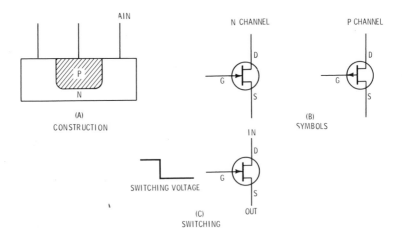

Figure 9-5.   FET Switching.

the silicon. The DC resistance is several hundreds of ohms. A deposit of P-type impurity along the top of the sheet will limit current through the sheet if this P-N junction is reverse-biased. The negative voltage on the P side of the junction and the relatively more-positive voltage on the N side cause charge carriers to be drawn away from the junction, leaving a region depleted of charge carriers which will not pass current. As negative voltage on this "gate" element is increased, electron charge carriers are drawn farther away from the junction, further constricting the channel width which remains conductive, until eventually a "pinch-off" voltage is reached. The symbol for an N-channel FET is shown in Figure 9-5B. The end of the channel which injects the negative charge carriers is called the source (comparable to the cathode of the vacuum tube), and the end which collects the charge carriers is termed the drain.

Switching with an FET can be accomplished between source and drain by applying a switching current to the gate as shown in Figure 9-5C. Certain characteristics of the FET as a switch are advantageous in some applications. The gate exhibits a high input impedance and does not load switching impulse circuitry. FET's may be made PNP or NPN. A PNP type FET acts like a vacuum tube, that is, the gate draws current when positive but not vice versa; the gate turns off when negative, and its characteristics resemble a vacuum tube. The NPN type cuts off with a positive gate but otherwise is the same.

## SEQUENTIAL SWITCHING

Many installations in hospitals, industrial operations, and security systems require that a single individual must monitor the output from several cameras. To attain this, electronic units known as sequential switchers (see Figure 9-6) cause the signal from remote cameras to be automatically displayed in sequence with an adjustable interval. Sequential switchers result in manpower savings by increasing surveillance capabilities.

In most sequential switchers the sequence is generated by a ring counter as shown in the schematic of Figure 9-7. A ring counter is a circuit consisting of a chain of monostable stages that mutually operate in succession. When the last in the chain has

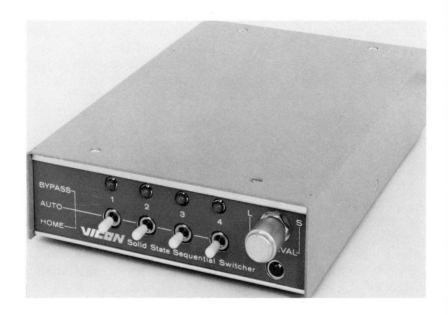

**Figure 9-6.**  Sequential Switcher.

been triggered and at the instant it returns to the dormant state, a pulse triggers the first stage in the chain, causing the crowd to begin again, thus a continuous counting is set up. To cause the stages to count at uniform intervals, each one is uniformly biased so that it will not by itself pulse back and forth, but must "wait" for an additional pulse from another separately supplied pulse. In the unit shown, this pulse is furnished by a continuously pulsing unijunction oscillator, and the pulse rate—hence, the sequence interval—may be adjusted. Each stage of the ring modulator initiates solid state switching of the video.

## SYNCHRONIZED SWITCHING

When the video signal is randomly switched on to the monitor from one camera to another, the picture "jumps" and a moment is required for the monitor to become synchronized with the

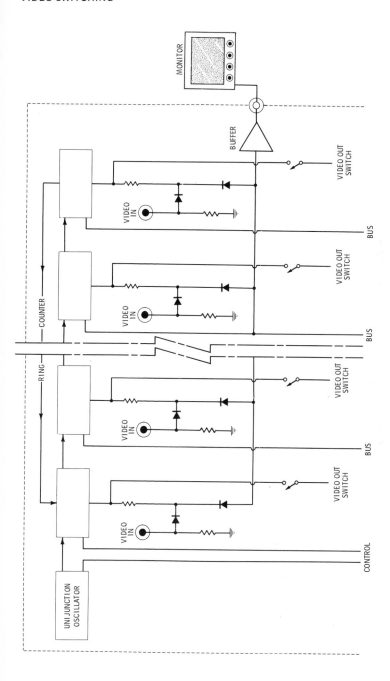

Figure 9-7. Sequential Switching Diagram. (Courtesy of Vicon Industries, Inc.)

new signal. Moreover, when a signal is fed to a video tape recorder, the nature of tape recording devices is such that the signal must be switched only during the brief intervals at the start of a new frame. In some video surveillance installations (and in all broadcast TV), all cameras, switches, and other equipment, a common sync is employed, however, it is possible to switch during the vertical sync interval. The technique is to feed the video signal to the switcher and strip off the vertical synch pulse. The leading edge of this pulse is then used to actuate a solid state switch. A *vertical interval switcher* is designed to operate in the manner just described.

Many cameras are now manufactured with adjustable vertical phase. This produces clean switching without the use of synchronization.

## TIME LAPSE SWITCHING

A great advantage of modern video surveillance is that a video tape of a given area can be made at intervals, thus permanently recording the identity of individuals or activities in a certain area. One method is to use a single tape recorder and to place a single frame alternately from a number of cameras. Certain later model cameras are designed especially for this function. A time lapse recording system is diagrammed in Figure 9-8. A further feature of the time lapse tape recorder is that the frame sequence may be separated and played back in such a manner that only the output of one camera is displayed. The frame speed at playback may be varied so that scenes may be selected and halted for closer inspection.

Still another advantage is that it is not necessary to record continuously, but frames may be taken at intervals of seconds or even minutes. By this means a continuous video record is available if desired, yet a tape may be erased and reused. Although, there is thus a great advantage in video tape surveillance over photographic surveillance. Video tape is also available for instant replay as often as desired until erased.

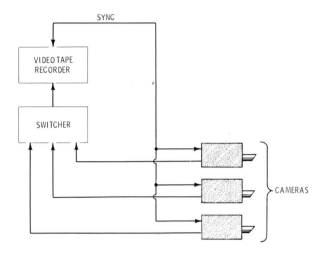

**Figure 9-8.** Time-Lapse (Synchronized) Switching.

# 10

# *Control of Accessory Equipment*

Remote control means that some form of signaling must be employed between the control point and remote unit. Not only must the pan and tilt of the camera be controlled, but such functions as lens adjustment, window wiping, and so on must be initiated. Signaling may be accomplished by a single pair of wires or even by employing the same coaxial cable that is used to transmit the video signal back to the control point.

The method of remote control of pan and tilt drives previously described is to provide electrical power from the control itself. Multiconductor cable is usually used for this, which mates to the units through connectors. Weatherproof connectors are used with outdoor units. A disadvantage of this method is the size of conductors required between control and remote drive. For example, one manufacturer of heavy duty 24-volt drives recommends the following wire sizes:

| Wire Size (AWG) | Maximum Distance (feet) |
|:---:|:---:|
| 22 | 350 |
| 20 | 550 |
| 18 | 900 |
| 16 | 1400 |
| 14 | 2300 |

In a pan and tilt drive with auto-scan, this would mean eight conductors of this size. Cost of wire and installation thus becomes considerable at longer distances.

One method of reducing wire size is to employ relays at the remote site and drive the unit from a local source of line current. In the same simple control system so far described, this still results in multiple conductors, however. Thus, there is a great advantage in employing a more sophisticated means of control using less wire. The saving in wire cost alone means that more cost can be directed to the control system itself.

## MODULATION

Conveying a signal means that some form of *carrier* must pass through a *medium*. By imposing *modulation* on the carrier, signals may be transmitted. When a pair of simple wires are used as a medium, pulses of current may be used for signaling. In this case, the carrier is the current and the pulses are the modulation. Circuitry may be employed at the control point to encode the pulses. Pulses are decoded at the remote point to initiate the desired action.

The idea of modulation is closely related to the transmittal of information, intelligence, or to put it another way, command. Modulation takes place when a form of energy changes state, that is, when something in its characteristics is altered in a detectable way. The most simple thing is to turn a current or energy source on (or off, if it is on). This is the simplest form of intelligence, or logic. That is, a state may either exist (known as Logic 1) or it may not exist (known as Logic 0). The conveyance of the knowledge of Logic 1 or Logic 0 is the simplest form of knowledge that it is possible for us to conceive. This is known as a *bit*.

Intelligence is made up of a nearly infinite combination of bits (see Figure 10-1).

The bandwidth of a system determines the number of bits per second that can be transmitted. We saw in discussing the television signal that if we desire higher picture resolution, more bandwidth is required. This is because a clearer picture requires that much more information regarding shades of gray, definition, and motion must be known at the receiving end, or more "bits" of intelligence must be transmitted. In summary, one bit consists of a single change of state of the transmitting carrier. The change need only be sufficient to be detected at the receiving end.

Any characteristic of a signaling carrier may be changed. In Figure 10-1, we made the simplest change that can be conceived by merely changing the level of a current from zero to another value. An alternating current may be changed in amplitude (AM) or frequency (FM), and so on. The type of modulation and the type of carrier determines the limits of the system in all respects, including economic limitations.

When longer distances are required, radio frequency may be employed as a carrier with either FM or AM. Microwaves may be used when free space transmission is employed or lower frequencies may be transmitted over coaxial cable. A disadvantage of coaxial cable is that higher frequencies attenuate greatly when this medium is employed. One method of control would be to use a carrier at slightly greater than video frequencies and use the same

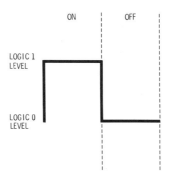

**Figure 10-1.** Logic and Bits.

**Table 10-1.**   Methods of Control

| Medium | Type of Modulation | Advantages | Disadvantages |
|--------|--------------------|------------|---------------|
| Wire | None | Simplicity of circuitry | Cost (heavy wires) |
| Wire | Pulses | Lighter wire (two conductors) | More complex circuitry required |
| Coaxial cable | Fm, a-m | May be combined with video | More complex circuitry required |
| Coaxial cable or wire | DC variation | Simplicity, may be combined with video | Coaxial cable must have stable dc resistance |
| Space (microwave) | Fm, a-m | Very long distance | Cost; system must comply with FCC regulations |

coaxial cable for both video and control. Another approach would be to go to the lower frequency as an extreme, that is, employ DC as a carrier (zero frequency) and make slight, slow going changes in current and voltage serve as modulation. We can tabulate all of the conceivable means of control as shown in Table 10-1.

## LONG DISTANCE CONTROL

This is a class of control in which the video signal is not merely coupled from camera to monitor, but one where distances are so great or cameras are so inaccessible that conventional simple hookup is not possible. In this case, microwave, leased lines, or special low loss cable is employed.

Ordinary telephone circuits may be employed to send control signals. One method would be to employ tone signals, known as Tone Control. A tuned circuit or active solid state filter can be employed to respond to a given tone which in turn triggers a relay for the desired function. Whenever conventional telephone circuits are used, the telephone company will require that the type of impulse to be sent must be approved so as not to interfere with the telephone system. Conventional telephone circuits respond to frequencies between about two hundred Hz to about 3000 Hz. This is sufficient bandwidth for a large number of functions to be controlled.

Telephone circuits are usually not designed to transmit direct current signals. When on-off digital pulses are used for command signals, it is necessary to use an audio frequency carrier that may be modulated with the pulses. Another method is to employ Frequency Shift Keying, which will be discussed later.

It is also possible to lease coaxial cable circuits over which both control and video signals may be sent. Command signals as well as audio can be included by adding a carrier of 4-5 MHz to the video signal which ordinarily does not extend above 4 MHz. Leasing costs for coaxial cable are very high, however, and cable is not widely available.

Microwave has a peculiar advantage over the simpler types of coaxial cable. It has a very wide bandwidth, making it possible to send more than one video signal over the same carrier as well as any combintion of control signals. When microwave is employed, an FCC license is required and custom engineering is usually necessary, even though "off-the-shelf" hardware is used. Microwave energy is concentrated in beams and travels only in line of sight paths. Parabolic or dish shaped microwave antennas must be sited and raised to form line of sight paths.

## FREQUENCY SHIFT KEYING

This form of modulation (called FSK) may be used to convey pulse modulation over several different carriers through different mediums. For example, it may pass over a pair of wires, through commercial telephone circuits, thence over microwave, or in any combination of the foregoing. FSK modulation is simply a combination of frequency modulation (FM) and pulse modulation (PM).

It consists of two distinct audio frequency tones, one called *mark*, or Logic 1, the other called *space*, or Logic 0. The two frequencies alternate. When FSK is applied to an audio frequency FM detector, the detector will have two DC outputs, one for mark, one for space. The pulses are "clocked" in the form of codes. That is, it operates the same as the primitive Morse Code, but because electronic circuitry can send and receive much faster than a human, a number of codes consisting of mark (dot) and space (dash) have been standardized for electronic transmission. In its most idealized form codes are based upon digital arithmetic combined with logic (Boolean) theory.


## PRESET CONTROL

Systems have been devised for driving the camera to a preset position in pan and tilt at the touch of a button. Such a system would be valuable in any surveillance installation which merits the additional cost. An observer seated at a monitor can easily lose camera orientation when it is necessary to pan the camera throughout its range. By automatically bringing the camera back to a prearranged point, surveillance control can then resume from that point. It may also be that there are several points which are of primary interest, from which a camera need only occasionally be moved. These points may require that a lens be zoomed and focused on a given scene. A further refinement, therefore, would be to also provide a means of setting the lens to a prearranged setting for the desired scene. Thus, with the touch of a button the camera would traverse to a point, zoom, and focus to a pre-arranged scene (see Figure 10-2).

Figure 10-2.   A Preset Position Control Unit.

Electromechanical devices for selective movement have been in use for many years, although they are only now coming into use for video surveillance.

The first requirement is that the pan and tilt drive be connected to the point of control with a sensing device. This device could be a *synchro*, or a potentiometer that has its moveable arm positioned by the pan or tilt mechanism. Once the instantaneous position of the pan and tilt drive is sensed electrically at the control point, electronic circuitry can be employed to regulate the driving motor and bring the unit to a prearranged setting.

One circuit for driving a camera or lens to a preset position incorporates a potentiometer in the lens or camera drive as previously mentioned. A voltage is then sent back to the control point which is compared to a preset voltage using an integrated-circuit operational amplifier as a *differential comparator*. IC voltage comparators in this form are now widely available. Refer to Figure 10-3.

Input voltage from the shaft potentiometer is compared to a preset level. If the input is more positive than the reference level, output of the comparator is positive. If input is more negative, output goes negative. A pair of drivers, one for negative and one for positive, now drives the shaft until shaft and reference voltages are equal. For lens functions using permanent–magnet DC motors, transistors may drive directly. Otherwise relays may be used to actuate AC motors or larger DC motors of the pan and tilt drive.

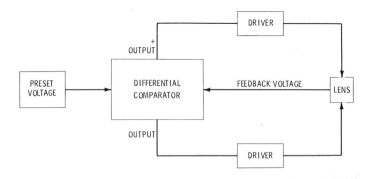

**Figure 10-3.** Differential Comparator.

The four reference potentiometers allow four preset positions to be selected. The potentiometers can be screwdriver settings made by watching the monitor and making the setting. Thereafter an attendant merely depresses a switch to attain the desired preset position.

## MISCELLANEOUS CONTROLS

Individual installations often make use of any number of different control arrangements. Usually custom layouts do not employ specialized electronic circuitry but are individualized arrangements of switches. Control from a central console may be shifted from one pan and tilt drive to another by a multipole switch.

It is possible to display the instantaneous position of a pan and tilt unit at the operating console by connecting a voltmeter to an energized potentiometer that is positioned by the drive shaft. The voltmeter may be calibrated in degrees of pan and tilt. When a unit is designed to autopan continuously between limits, it is possible to devise circuitry to vary the limits of the autopan sector from a remote location. This can be done using voltage comparators with circuitry similar to that previously described for the preset position control.

# 11

# *Signal Processing,*
# *Special Effects and Color*

Our purpose now is to put together many of the systems and processes found in video surveillance. The television camera with its refinements can be made to see in semi-darkness or observe infra–red images in total darkness. It can see colors and detect motion. The monitor screen can be broken up or split to present more than one image. The CCTV industry is growing and many new ideas and methods are coming into use to satisfy special needs. In this chapter we will examine some important techniques, beginning with video processing.

## SIGNAL PROCESSING

It is often necessary to process the video signal in order to make up cable and switching losses, suppress hum, or perform other operations upon a signal. This is ordinarily grouped under the heading of signal processing. In simple installations little processing is necessary. As switches and other devices are added, however, processing sometimes becomes necessary. Losses incurred in switching and cable runs may be overcome by amplification, hum reduced by isolation or other means, and carrier modulation employed for transmission.

Processing equipment now offered to the industry is completely solid state with more and more use made of IC's. It is usually made to fit standard Electrical Industry Association (EIA) racks measuring 19 inches wide for console mounting. Larger installations employ standard large cabinet racks in control centers.


## Amplification

The standard monochrome video signal is specified as one-volt peak-to-peak amplitude, occupying a frequency band from about 60 Hz (the field rate) to about 8-10 MHz. This idealized form is rarely attained, although even with considerable degradation many systems operate satisfactorily. There is a point, however, where degradation cannot be tolerated.

A video amplifier can do two things: it can reproduce a signal in amplified form, and it can enhance those portions of the signal spectrum that have been attenuated. Amplifiers are sometimes traditionally classified as *voltage* amplifiers or power (also referred to as *current*) amplifiers. That is, when comparing input to output, if there is a large voltage gain, the term voltage amplifier is used. When there is little difference in voltage but still a large power gain from input to output, it is called current amplifier. When dealing with the low impedances associated with coaxial cable, power amplification is the most useful parameter.


## Distribution Amplifiers

It has been pointed out that coaxial cable must be terminated in its characteristic impedance. In most cases, several high impedance devices may be placed across the output when it is terminated only at the end device. All bridging, together with the termination, must be done at approximately the same physical location, however. In some cases this is not possible. For example, if one or more monitors or a video tape recorder is to be separately located, all devices must be fed by coaxial cable.

Figure 11-1. Distribution Amplifier. (Courtesy of Pelco.)

A distribution amplifier may be employed for this. The signal to be distributed is terminated at the distribution amplifier. The distribution amplifier consists of several simple amplifier circuits in parallel which are fed by the signal. Each amplifier then feeds a separate coaxial cable to the respective terminal location. Each of these terminal locations contain a separate terminating impedance. If the terminating device does not have a matching terminal impedance, a resistor is used to arrive at the correct value. Figure 11-1 is a photo of the Pelco Model DA-104 Distribution Amplifier. No adjustments are required. This unit has one 75 ohm input and four separate outputs, each of which may be applied to a separate cable with 75 ohm impedance.

## Equalization

An equalizer is a device for selectively amplifying certain frequency components of a signal. When a current of certain frequency passes through a line, it undergoes both attenuation and phase shift. The television signal is made up of many frequency components, and with each scene and movement the proportion of components at each frequency changes. Moreover, the amount of attenuation and phase shift will be different for each frequency component. In monochrome television, phase shift is relatively unimportant, but relative attenuation of frequency components will cause picture degradation. The use of equalization will reasonably bring up most components to the same level.

Figure 11-2B shows the attenuation characteristic curve of a line. An ideal equalization system would have a response curve that is an exact complement, as shown in Figure 11-2A. Attenuation due to frequency is sometimes known as roll-off. An equalizer will usually have a means of adjusting to the roll-off characteristic of a given system. This is done by providing a means of inserting gain and adjusting the slope. An amplifying stage may be made with variable frequency response as shown in Figure 11-3. The emitter resistor offers degeneration at lower frequencies where the capacitive reactance of capacitor C is high.

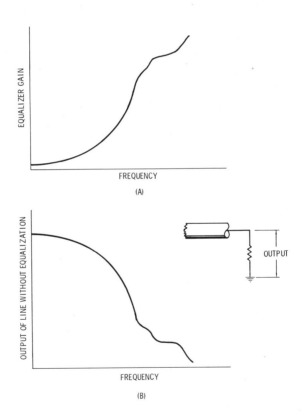

Figure 11-2. Idealized Equalization. (A) Equalizer Gain Versus Frequency. (B) Attenuation Without Equalization Versus Frequency.

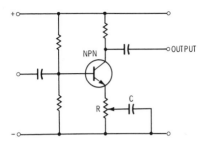

**Figure 11-3.** Equalizing Amplifier.

The capacitor shunts higher frequency components past the resistance depending upon the setting of potentiometer R. The potentiometer is set to compensate correctly for the amount of roll-off present. A simple circuit such as this is usually satisfactory for use at the end of a line carrying monochrome signals. More sophisticated methods are available that can develop a response curve that is a good complement to the roll-off.

## Hum Reduction

When using an unbalanced line, especially at longer distances and where different AC power circuits are used between the driving and terminal end, there is a good chance that a difference in potential exists between cable shield and equipment chassis. This is sometimes called a "hot" cable. Since the shield is half of the signal path, the stray AC current in the shield becomes mixed with the signal. Ground resistance and cable length are the factors that cause hum. This is the reason for using a good ground connection at both ends of the line. A good ground has the effect of shunting the resistance of the cable shield.

Isolating the terminal end from ground will reduce hum, but the method is dangerous since large potentials may develop. Another method is to employ a clamper amplifier. A clamper uses the horizontal sync from the video signal to establish a constant voltage reference point. Since this process occurs at a 15,750-Hz rate, the reference will be reestablished over 260 times during one cycle of an interfering 60-Hz hum signal. When using a clamper, all circuits preceding the clamp stage must linearly pass the hum and video, otherwise the video will be degraded. This places a limit of about 1 volt or less of hum on a 1.4-volt video signal.

Hum as well as extraneous induced noise reduction can best be accomplished by the use of a balanced line with differential amplification. A differential amplifier is now available as an integrated-circuit chip. The differential amplifier is characterized by two input terminals, neither one of which is necessarily grounded, and two output terminals with the output waveform of one being equal to but inverted with respect to the other (see Figure 11-4). The input voltage is considered to be the difference in potential between the two input terminals, and the output is taken as the *difference* between the two output terminals. The differential amplifier is a balanced amplifier, and tends to compensate for inherent drift. In a balanced transmission line, hum and induced noise will be common to both terminals with reference to ground. This is known as a *common mode* component. The differential amplifier will reject the common-mode signal and amplify only the difference signal, which is the video signal, between the two output terminals. Common-mode rejection is an important parameter in this type of transmission scheme. The *common mode rejection ratio* is the ratio of the desired transmitted signal and undesired stray voltages that are common to both conductors of the transmission line.

## SPECIAL EFFECTS

### Character Displays

There is an increasing demand for the display of numerals or alphabet characters on the monitor screen, known as alph-numeric display (see Figure 11-5). One use of character display is to indicate the time and date. When the scene is recorded, a positive record is then available. Another use is to record the camera

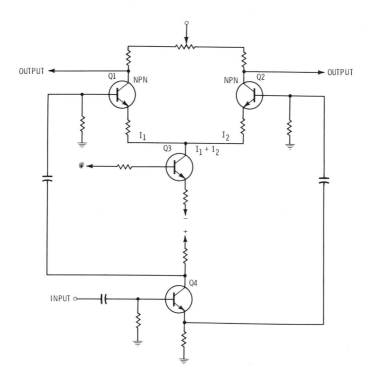

**Figure 11-4.** Differential Amplifier Driver for Balanced Line.

number. Such video records then become evidence, and many courts are beginning to accept this form.

Characters are built up from a modulus as shown in Figure 11-6A and 11-6B. It can be seen from Figure 11-6A that any numeral can be built from fifteen squares, with the figure 8 using the maximum configuration. An alphabet character can be built in the same way. Logic circuitry is used to produce the waveforms to be superimposed on the video signal. With reference to Figure 11-7, note the waveforms necessary to produce numeral 5. To be completely useful it should be possible to move the character anywhere on the screen. Another feature is to provide an automatic brightness control so that the character will provide just the right brightness relative to the scene background at the

**Figure 11-5.** Alpha-numeric Character Generator. (Courtesy of
Telemation, Inc.)

point of display. Characters should be stable and flicker-free.
Because many CCTV installations use random interlace, a character display generator should be compatible with this type of
scanning as well as 2:1 interlace.

A number of methods have been used to develop the character
waveform. At least one manufacturer produces an integrated-
circuit character generator. One approach is to consider the
character display area as consisting of fifteen squares as in Figure
11-8. Each square may be defined by two time intervals, horizontal and vertical. The vertical interval is with reference to the
vertical sync and the horizontal with horizontal sync. We may now
generate two pulses. When the two pulses are in time coincidence,
we trigger another pulse which is superimposed on the video.
This would form a bright square. The numeral 2 would then consist of squares A,B,C,F,I,H,G,J,M and O. The numeral 1 would be
B,E,H,K and N, and so on. A pulse delay or shifting network
would be used to move the character on the face of the screen.

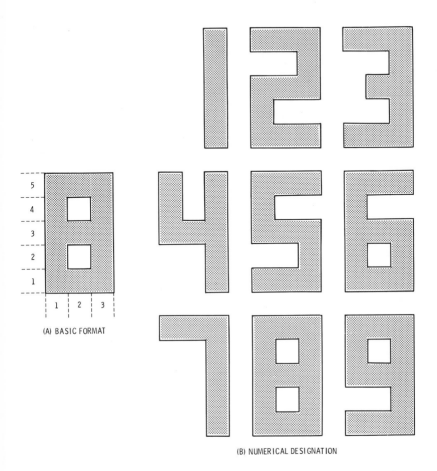

(A) BASIC FORMAT

(B) NUMERICAL DESIGNATION

**Figure 11-6.** Character Generation Modulus.

## Screen Splitting

There are applications wherein we may desire to place two or more pictures on a single monitor screen. The only limitation to this is the resolution of the system in TV lines relative to the detail required in each image. One application is in banking or retail sales, when it is desired to display a check along with an individual's picture. Some systems use an optical splitting method

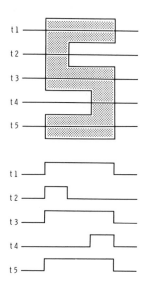

**Figure 11-7.**   Numeral 5 Waveforms.

**Figure 11-8.**   Forming Numeral From Squares.

with lenses and prisms, as diagrammed in Figure 11-9. Electronic methods appear less expensive in the long run, however.

Methods of electronic screen splitting are somewhat related to character generation. We first generate a pulse related to the time that is occupied by one area screen to be split. This pulse is then used to operate a diode or other solid state switching arrangement. The switch in turn switches from one video signal to another in such a way that the video images share the screen. When screen splitting is employed, of course, all video concerned must be in common sync.

The block diagram in Figure 11-10 shows a method of screen splitting. In this system the screen splitter is a separate unit and splits the screen into two images. The video from two cameras is fed in and one composite signal then goes to a monitor. Sync is separated and a vertical and horizontal reference is derived. Each then goes to form a switching pulse. The vertical switching pulse is long relative to the horizontal switching pulse and represents the time that the raster is scanning through the number of horizontal lines that will go to make up one of the split images. The horizontal switching pulse represents the interval that the horizontal scanning line will spend on one of the split images.

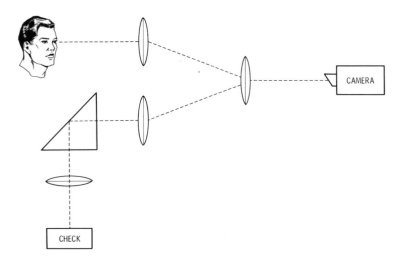

**Figure 11-9.** Optical Splitting.

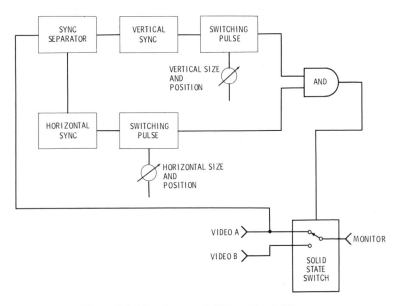

**Figure 11-10.** Screen–Splitting Block Diagram.

The size and aspect ratio of the split images depends entirely on the switching pulse lengths and will be made adjustable.

When both horizontal and vertical pulses are coincident, one of the images must be scanned. The AND logic circuit determines this. During this time the video switch will be in position to incorporate one or the other image into the composite video signal which goes to the monitor (see Figure 11-11).

To summarize screen splitting, one may point out again that the process depends entirely upon resolution. With only 390 lines available for the actual image in a standard 525 line system, two images will have only about 180 lines of resolution if the screen is split exactly in half.

## IMAGE MOTION DETECTION

Motion detectors in a video system serve as intrusion alarms. When a camera is on a fixed scene in which no motion takes place, the video signal will have a steady output that is repeated

**Figure 11-11.** Video Security System Use of Screen Splitter. (Courtesy of Thalner Electronics Laboratories, Inc.)

continuously. If motion then takes place within the area covered by the camera, however, the video signal will not repeat itself but will deliver a varying amount of current through a resistor and diode combination. A logic circuit can be designed which will cause an alarm whenever the video level changes due to motion. The circuit may be designed so that it will alarm whenever the average video of one frame varies relative to the other. A good method is to sample the video every few frames and compare the samples. To insure adequate surveillance with video motion detection, a sensitivity adjustment is usually employed which is manually set for a given situation.

When using a video motion detector that responds to motion throughout the scene, certain precautions are necessary to prevent false alarms. Random movement in the scene can cause signal change. Examples of this would be objects moving in the wind or an interior camera scene which also includes a window, outside of which motion appears. In an outdoor situation other factors which may cause false alarm are rapid cloud changes, sun shift causing comparatively rapid reflection changes, or the presence of moving lights outside of the field of view.

Another factor is camera location and direction. In general, indoor and night time outdoor situations are the simplest to monitor with video motion detection. Judgment must be exercised in selecting the camera lens combination, setting, and position so that actual intrusion will be clearly distinguished. Focal length and camera setting should completely cover the exact area to be guarded, and camera sensitivity range should be compatible with the scene and prevailing light.

Once field of view and camera position are established, the sensitivity setting of the motion detector becomes important. For example, if an opening door is selected as the alarm, sensitivity should include the change in signal caused by the door opening, but no less. Conversely, when minor changes in the scene are brought about by an anticipated intrusion, the sensitivity setting must be greater. In summary, therefore, two factors govern the installation and operation of a video motion detector:

1. Camera characteristics, location, direction, and field of view.
2. The sensitivity setting (which may be manually varied by operating personnel to adapt to periodic conditions).

Certain electrical factors can also cause false alarms. Since the motion detector operates on signal changes, the same effect can be duplicated by fluctuations in line voltage energizing the camera or any other transient electrical condition. The video motion detector itself should be designed to eliminate false alarms due to respective line voltage variation and slowly changing video conditions. The camera may not be so protected, however, or the camera may also be faulty, causing a fluctuating signal. Faulty connections or transmission interference are also a cause of false alarm.

One approach to video motion detection is to employ screen splitting techniques and set the motion detector up so that only motion in a certain sector will cause alarm. The chosen sector may include doorways, stairs, or other intrusion points. Another approach is to employ photo electric pickups on the face of the monitor itself. On one such model now available, suction cups containing two photo-electric transducers are placed at certain points on the monitor screen. Light changes at that point on the monitor screen then cause an alarm.

It is important to realize that a video motion detector is in fact an alarm device. It is not a surveillance device, as defined in the chapter on video security analysis. It is, therefore, an actual step backward in the video security process, and assumes that the protected monitor is not being observed. Video motion detectors operate on the principle that some unique element in the video signal undergoes a change when intrusion occurs. It is theoretically possible, using computer technology, to design a motion detector that responds to certain selected, very unique elements in a video system. Practically speaking, this may be accomplished more efficiently by using sonic, microwave, or other alarm devices. The system designer must take this into consideration when designing specific installations.

## COLOR

In the majority of video surveillance applications it is not necessary to employ color. The reason for this is that color equipment is more expensive to install and maintain and little advantage is gained through its use. Nevertheless, there are a few situations where color is used in CCTV security monitoring, and this discussion will aid in understanding its principles.

Light as perceived by the human eye is made up of wavelike radiant energy that exhibits a very short wavelength such that there are from 36,000 to 63,000 waves per inch. What we perceive as white light is actually a mixture of waves of different length, or radiant energy of different frequencies. Lower frequency or longer wavelength light is red, and higher frequencies tend toward the blue or violet side. As we progress through the range of light from lower to higher frequencies the colors vary to form the spectrum. The color spectrum is shown in Figure 11-12. Colors that appear to us in ordinary viewing do not consist of light of a single

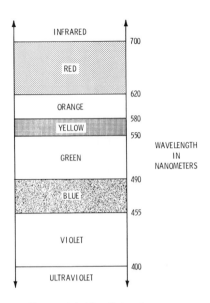

**Figure 11-12.** Color Spectrum.

frequency. Rather, what appears as a certain color may be made up of light energy of a band of frequencies with a predominant center frequency. Scientists have discovered that it is possible to synthesize any color by choosing three so-called primary colors from points within the spectrum, then adding white (the entire spectrum) and mixing quantities of these three chosen primaries.

The three primary colors are blue, green, and red. With these three colors and the addition of white light, any color can be formed. It has been found that there are two parameters to a color: *luminance* and *chrominance*. Luminance refers to the amount of white light and chrominance is the amount of colors or hue present in a given light sensation. In order to send a color picture, it is thus necessary to send a signal conveying the luminance and chrominance for each point on the scene to be transmitted. However, one element, luminance, is already being transmitted in ordinary black and white (monochrome) television. It is therefore only necessary to add chrominance information to the modulation in order to create a color picture. This chrominance signal will then tell the receiver the proportion of red, green, and blue to be mixed for the color at any given instant. The proportion of white light determines the shade of color. For example, pink is merely red with larger white content. The color image tube has three outputs, one for each of the primary colors. Methods of scanning and synchronization are set to the same standard as they are with monochrome TV. From the three primary colors, the camera creates a luminance signal and chrominance signal.

When color is used in CCTV, the important thing is that the signal must be transmitted withou degradation of the various frequency components. Color is picked by the camera and displayed on the monitor by tiny areas of the screen which are sensitive respectively to the three primary colors. For every instant during the scan, each small color area of the screen must respond to the correct amount of primary color so that the resultant will be the correct proportion of red, blue, and green to produce either pure white or the infinite combinations that make up true colors. This means that the color signal must be properly transmitted in both phase and amplitude.

## Analyzing the Color Signal

Color signals are derived by varying the phase and amplitude of a modulation component relative to a reference. The reference is called a "burst," or short pulse of 3.58 MHz. Chrominance is a function of *phase*, and the luminance is a function of *amplitude*.

There are three parts to the color TV signal, known as Y, Q, and I. The Y portion of the signal is the luminance portion. It carries sufficient information to provide a black and white picture, and it is formed at the camera by taking 11 percent of the blue output, 59 percent of the green, and 30 percent of the red signal output and combining them. The choice of percentage is a standard and relates to the properties of light as well as pickup tube characteristics, etc. The Y signal may be displayed on an ordinary black and white monitor. It has the usual bandwidth of from 50 Hz to about 4 Hz.

The chrominance signal is formed by combining Q and I, and these in turn are formed by mixing certain percentages of the primary color voltages from the camera, all as shown in Figure 11-13.

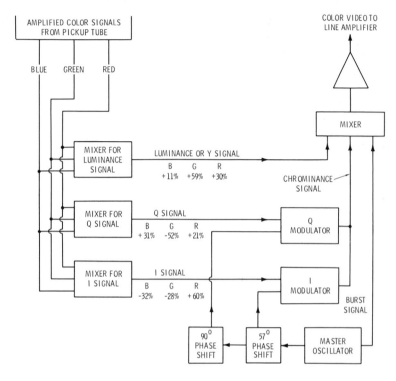

**Figure 11-13.** Generator of Color Video Signal.

With reference to Figure 11-13, note that the Q signal consists of 31 percent of the blue camera voltage, 52 percent of the green, and 21 percent of the red. The blue and red components of the Q signal are plus voltages, while the green is a minus voltage. A minus signal is one whose voltage changes are inverted or are in opposite phase with respect to the plus signal. The I signal consists of minus blue and green and plus red in the percentages shown. The Q signal and I signal go to modulators and modulate a 3.58 MHz carrier. The carrier is shifted 57° in phase and fed to the I modulator, then shifted another 90° and applied to the Q modulator. The I and Q signals become modulation on the 3.58 MHz carrier, then are combined with the luminance signal to complete the color video signal. Although I and Q signals are modulated onto the 3.58 MHz carrier, the 3.58 MHz frequency is suppressed at the camera and only the modulation sidebands go out in the signal.

The sync signals are the same as with standard monochrome, but with the addition of a color burst signal. This is a burst of 8 cycles from the 3.58 MHz carrier which was otherwise suppressed and occurs during each horizontal sync period.

In the color monitor are two demodulators that recover I and Q signals, by combining chrominance signals with a 3.58 MHz demodulating signal. The demodulating signal has been phase-locked to the eight cycle burst, which is a sample of the 3.58 MHz carrier from the color oscillator at the camera. The monitor color oscillator must remain locked in phase with the color carrier oscillator. Automatic phase control circuitry is used for this. When this locally generated 3.58 MHz signal is applied to the color demodulators, we recover the I and Q signals shown on the color phase of Figure 11-14. The positive I signal (+1) lags the reference phase by 57° and the positive Q signal (+ Q) lags another 90°. Negative I and Q signals (—I and —Q) are inverted or differ by 180° from their positive counterparts. Vectors in the figure show how a red color voltage may be secured by suitably combining 60 units of + I and 21 units of + Q signals. The red vector lags the reference phase by 76.5°, as in Figure 11-15. A green color voltage would result from taking 28 units of the —I and 52 units of the —Q signals. Blue would result from 32 units of —I combined with 31 units of + Q.

Another approach is to use R—Y and B—Y signals, rather than R and B. The resultant vector is illuatrated by Figure 11-16. In this case the demodulator is supplied with a 3.58 MHz signal, which produces an output lagging the reference or burst phase

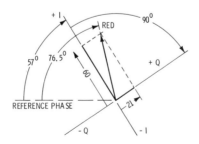

Figure 11-14. Red Color Phase Resulting From Combination Of I- and Q-Signals.

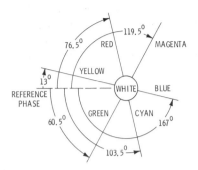

**Figure 11-15.** Phase Angles of the Color Signal With Reference to the 3.58-MHz Color Burst Signal.

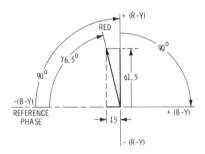

**Figure 11-16.** A Second Method of Deriving Red Color Phase.

by 90°. This output is not the I signal, instead it is a "red-minus-Y" signal. indicated by the symbol R—Y. Inversion produces a similar negative signal, so that there are positive and negative R—Y signals, usually shown by the symbols +(R—Y) and —(R—Y).

The other demodulator no longer delivers positive or negative Q-signals, but delivers "blue-minus-Y" signals, marked B—Y. The positive B—Y signal is shown as +(B—Y) and the negative as —(B—Y). Note that —(B—Y) is in phase with the burst or reference voltage.

The "Y" which is referred to in all these symbols is the luminance signal. As an example, the R—Y signal may be thought of as red with the Y signal lacking. If the Y signal is added to R—Y only the red color voltage will remain. The R—Y signal represents this combination of color voltages:

$$+(R—Y) — +70R —59G —11B$$

A positive Y signal consists of the following:

$$+Y — +30T +50G +11B$$

Combining these two signals gives a total of +100 red, since positive and negative values cancel for green and blue, thus leaving only red. Similarly, combining +Y with +(B—Y) leaves only blue. Likewise, there is developed a green-minus-Y (G—Y) signal. Combining +Y with +(G—Y) leaves only a green color voltage. Combinations such as these are carried out in the matrixing system of the receiver.

When demodulator outputs are R—Y and B—Y, it is possible to combine certain percentages of these outputs to produce the same color voltages as with I and Q outputs. Figure 11-15 show vectorially that combining 61.5 units of +(R—Y) with 15 units of —(B—Y) will produce a red color voltage lagging the reference phase by 76.5°, just as when red is produced by fractions of I and Q signals. For green it would be necessary to take 51.5 units of —(R—Y) and 29 units of —(B—Y). Blue would result from 10 units of —(R—Y) and 44 units of +(B—Y). Phase angles for green and blue will be the same as when demodulating I and Q signals.

After demodulation, the I and Q or R—Y and B—Y signals are applied to matrixing circuitry. A method of matrixing demodulated I and Q signals is illustrated by Figure 11-17. The Q demodulator is here assumed to deliver a positive Q signal, while the I demodulator delivers a negative I signal. These signals go to phase splitters, which furnish both inverted and in-phase signals. Relative values of red, green, and blue for all the I and Q signals are shown. From the luminance or Y amplifier comes a positive Y signal. The signals go to matrix circuits which combine the proportions of Y, I, and Q signals. The combinations are red, green, and blue color signal voltages. The green matrix accepts 100% of the +Y signal, 64% of the —Q signal, and 28%

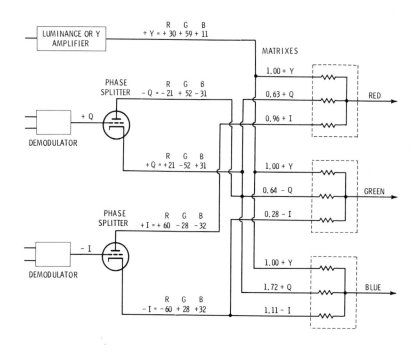

**Figure 11-17. A Matrix System.**

of the −1 signal. Adding the percentages of red, green, and blue (algebraically) and rounding off decimal fractions show that red and blue values cancel, while green becomes 100%.

| | | | |
|---|---|---|---|
| +Y × 1.00 − | +30R | + 59G | +11B |
| −Q × 0.64 − | −13R | + 33G | −20B |
| −I × 0.28 − | −17R | + 8G | + 9B |
| Sums | − Zero R | +100G | Zero B |

The combinations is the red and blue matrixes are related in similar fashion. The matrixed signals are applied to the respective red, green, and glue of a color tube.

## Frequency Components of Color Signals

It can be seen from the foregoing that color signals contain many complex waveshapes, even though they are contained in the same bandwidth as monochrome signals.

The relationship of sidebands in the Y signal and the I and Q components of the chrominance signal is shown in Figure 11-18. The luminance or Y signal is the equivalent of the video signal for black-and-white transmission and includes frequencies up to about 4.1 MHz. The Q component of the chrominance signal has sidebands extending both ways from 3.58 MHz for 0.5 or 0.6 MHz on one side and to about 1.5 MHz on the other side.

Parts of the video frequency range used for the luminance signal are used also for the chrominance signal, yet these two signals must remain separate during transmission and reception. Separation is possible because neither signal extends continuously throughout the video range, but consists of separated concentrations of amplitude around harmonics of the horizontal line frequency.

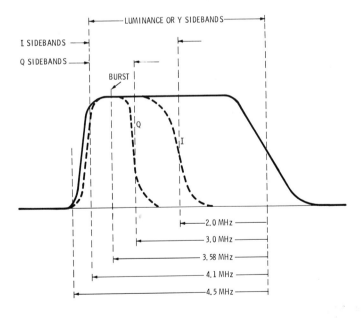

Figure 11-18. Frequency Spectrum of Color Video Signals.

The horizontal line frequency for color television is 15,734.26 cycles per second instead of the 15,750 cycles used for black-and-white. Luminance signal sidebands concentrate around harmonics of the color line frequency, but because the color subcarrier is an odd multiple of half the line frequency, concentrations of chrominance sidebands are shifted to half way between luminance concentrations. The color subcarrier frequency falls midway between the 227th and 228th harmonic of 15,734.26 Hz. This "interleaving" of luminance and chrominance signals allows both to be within the same video frequency range without interfering with each other (see Figure 11-19).

### Problems in Color Transmission

It can be seen from the foregoing that the 3.58 MHz burst is an important element of the color signal. Changes in phase of the burst during transmission, relative to the other frequency components of the signal, will result in errors in hue. Changes in relative

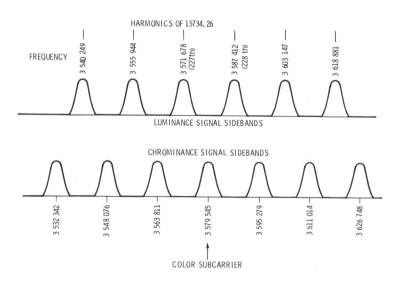

Figure 11-19. Interleaving of Luminance and Chrominance Signals in Color Video Transmission.

amplitude of the burst will result in color saturation errors—colors will be too vivid if relative amplitude is too high and vice versa. Phase errors are known as differential phase. The National Television Standards Committee requires a phase tolerance of ±3 degrees. More than this is noticeable at the receiving monitor when yellows tend toward olive or orange, depending upon direction of shift. Amplitude errors of the burst are known as differential gain and are seen either as washed out or as excessively glowing color. The Committee tolerance on differential gain is ±5 db, or 6 percent. In the overall system, each element in the transmission path, such as amplifiers, repeaters, or equalizers as well as the line itself, contributes to errors.

Our discussion of color has been centered around an analysis of the color signal itself. The reason for this is that in CCTV, color quality due to signal processing and transmission is the obligation of the user, rather than the public broadcaster. Color cameras and color video tape recorders are now coming into wide use in non-public areas. Equipment is necessarily more complicated than with monochrome, but the use of solid state integrated circuitry, nevertheless, is making equipment light, practical, and trouble free. The use of color in video surveillance is not widespread, but no doubt will grow. The choice of color depends upon economics.

# 12

# Procuring and Maintaining Video Security Systems

The video security industry began, when small, inexpensive cameras and monitors came on the market. Manufacturers of these items sought to create markets that would sustain mass production. This continues to be the case. Closed circuit television equipment has many other uses besides security. Camera and monitor producers satisfy all markets in order to offer lower cost mass produced equipment.

Another industry has arisen to produce accessory equipment such as pan and tilt devices, switchers, and so on. These are produced in standardized lines that allow selection for individual requirements. Many of these manufacturers will also customize accessories for special installations that can afford the cost.

Effective procurement and maintenance, therefore, requires attention to sources of supply, equipment reliability, maintenance, and the economics of first cost versus operating cost.

## MAINTENANCE

Sources of trouble in a video security system can be considered according to probability of occurrence. The most likely source of trouble in any system will be in the mechanical devices. This would be the pan and tilt units or any other device involving mechanical motion. The next likely source of trouble is cables and connectors. It is particularly true when cables are exposed to abrasion and movement, or units are subjected to frequent connecting and disconnecting. The last suspected source of trouble would be in solid state electronic circuitry, particularly when it is subjected to vibration, movement, or temperature and other environmental extremes. Thus, a simple video system consisting of only cameras and monitors has been known to operate trouble-free for years. This can only be the case, however, if good design and quality components are used. Poor quality can make electronic equipment a primary source of trouble.

A pan and tilt unit mounted on the exterior of a building can be a very expensive maintenance item. Extremes of weather may prevent troublefree operation over reasonable periods. This requires maintenance personnel to climb or otherwise gain access to the unit for repair. If repairs are not immediately made, security is lost. This factor may influence the location of cameras. It may be more desirable in the long run to employ more cameras to obtain the same coverage that could be had by one less accessible unit. Many expensive video installations have fallen into disuse because of high maintenance costs.

## CHARACTERISTICS OF ELECTRONIC EQUIPMENT

The design of electronic equipment and electronically controlled devices has evolved to the point where most products can be expected to be of a conventional configuration. Electronic circuits are contained on printed-circuit (PC)—sometimes called printed-wiring (PW)—boards. In an electronically controlled device, such as a video tape recorder or motorized lens, wires or cables then radiate to the various actuating and sensory devices. There may be one or more of these boards in a unit. In some cases these are of differing sizes and interspersed among the various mechanical subassemblies.

PC boards may also be designed as plug-in unit assemblies. The purpose of a plug-in board is for ease of removal in trouble-shooting and repair. The use of these boards requires more expense in materials and assembly as well as design. Nevertheless, the advantage of easy removal is obvious. The advantage is lost, however, if a spare plug-in board is not immediately available. If the user repairs his own equipment, spare PC boards should either be purchased, or there should be some assurance that a replacement board can be obtained. It should be noted that a PC board, as a discrete component, must be entirely compatible with the make, model, and serial number of a piece of equipment. Manufacturers occasionally tend to modify electronic circuitry without attention to units that are already in the hands of users. A service organization must be aware of this.

The repair of printed-circuit assemblies, whether plug-in or otherwise, must be attempted only when there is some assurance of success. The first problem is the isolation of a defective compo-nent. A burned out resistor may be defected by discoloration, however the location of a defective "chip" or integrated-circuit may be difficult. Very often, PC boards which are returned to the factory for repair are found impossible to economically repair simply because the trouble is too difficult to isolate. Moreover, the removal of solder from delicate devices requires care and skill, as well as special tools. Once a defective component is located, it must be exactly replaced with one of the thousands of integrated circuit types now in use. Before attempting repair, the type of IC must be known and available.

## RELIABILITY ANALYSIS

It is possible to scientifically analyze failure probability of specific systems. Much study has been devoted to this. Although it is usually unnecessary to treat video security systems with involved analysis, the basic principle of reliability analysis can be extended to this. The process can be approached by considering the prob-ability of failure of given items over a period of time. The assump-tion is that all components fail given sufficient time.

Operating environment is an important independent variable in determining reliability. Environment may be systematically analyzed by considering the following stresses:

- Mechanical
- Temperature
- Electrical
- Other

These can be separately analyzed.

*Mechanical stress* comes through static force, shock, vibration, and air pressure. A static force is exerted by the means of mounting. It may lead to effects on those other categories of shock and vibration. These latter are interrelated. Mechanical shock comes about from inertia or pressure. Equipment is exposed to shock, twisting, and so on during installation and servicing. Operation of equipment is a series of routine shocks. It is well to remember that even the mere removal of a connector is in fact a mechanical shock, even though made negligible through good design. Vibration is a series of shocks delivered at a certain frequency. Vibratory stress can only be treated by the designer and installer. Semiconductor devices are hermetically sealed, but the seal may fail. Electrolytic capacitors may also be pressure sensitive. Pressure, therefore, is a stress that may only be considered by the designer or manufacturer.

The subject of mechanical stress, specifically shock, is important to the user, maintainer, and installer as well as designer. We speak of "wear and tear," which is, in fact, the many shocks that may occur routinely. Shock is taken here in its broad sense, meaning any momentary mechanical stress other than that brought about in the specific operation of a piece of equipment. It has been mentioned that the mere removal of a connector is, in fact, a shock. It is possible that the majority of failures occur from shocks taken in this broad sense. For this reason, the troubleshooter should consider mechanical failure first. This would mean inspection of cables, joints, connectors, printed–circuit plug–ins, and so on.

The next environmental category is *temperature*, which has been briefly mentioned earlier. Temperature stress may be static, shock, or cycling. Static temperature stress occurs from long periods of sun exposure during operation. High static temperature stress may result in electrical component failure. Plastic materials may have accelerated wear out. The effect of contaminants is accelerated. Cold causes brittleness. Lubrication is affected.

Temperature shock means sudden temperature change in either direction. Acceleration from hot to cold may result in mechanical breakage. Cycling from hot to cold and back results in thermal fatigue. Electrical components are most likely affected by thermal shock.

*Electrical stresses* are of four categories: voltage, current, continuous power, and cycled power. Voltage may exist with no current flow, but not vice versa. Voltage stresses may cause minute ionic (chemical) action such as corrosion. In semiconductor logic circuitry, this could result in failure. More gross corrosion may also occur in connectors. The effect of contaminants is accelerated by current flow. The effect of current stress is through the creation of magnetic lines of force, which in turn affect circuit action or induce corrosive voltages. Although both voltage and current stress have very illusive effects, power stresses are more obvious. Power produces heat which leads to temperature stress. Also, power applied to mechanical devices has obvious effect. The designer allows for this in rating the various system elements. Continuously applied power is the end usage of any system. It is the factor that leads to any failure due to normal use.

Cycled power comes about in start–up and stop as well as transient conditions. Transients are momentary surges in either direction produced by a variety of causes. The designer and installer must be aware of transient conditions that may lead to failure. When failure occurs, it will be in the component that is weakest at that moment.

*Other stresses* leading to failure are under such headings as humidity, corrosion, abrasion, and so on. The operator, maintenance technician, installer, and manufacturer all exert influence on these stresses. Abrasion, for example, may occur during manufacture as well as in operation. Again, we are considering each category and subcategory of stress as applying to every individual component or element of the system.

Contributions to failure begin with the raw material itself and continue through final design, manufacture, and installation. Actual failure, however, occurs in the field. No device is absolutely failure proof. Rather, we approach the problem by attempting to arrive at the reasonable time that a failure may be expected. Reliability may be increased at greater expense, but the relationship is not linear. We reach a point where further expense is unwarranted.

Reliability may be defined as the probability that a device or system will operate under specified conditions within specified limits of performance for a given time. Note that, strictly speaking, it is not necessary for the unit to become totally inoperative. The system is assumed to have failed when any part becomes inoperative or falls below performance specifications. There are many practical considerations to reliability.

A system may exhibit susceptibility to abusive manipulation that is apart from in-service use. The act of maintenance itself may in turn create stresses that lead to failures in a system that otherwise would not fail for a long time.

One way to measure reliability would be through actual operational test. A problem here is that, for statistical accuracy, a great number of systems must be tested for a long time. Another method is to increase stress for a certain operational time on the theory that the increased stress will relate to in-service reliability.

Most systems will at least have been tested to some extent through actual operation, even though the tests are not statistically adequate. To be statistically adequate, testing costs would be enormous. There is a condition known as "infant mortality," whereby it is assumed that many failures occur during the initial operation of a new system. On this basis, a newly built system can be run for a period before initial use, sometimes at increased stress. This is called burn-in.

The study of reliability is an acturial study based upon probability theory. It is a mathematical treatment stating that all systems will fail sometime, with the probability of failure increasing as the system operates longer. We have seen, however, that environmental stresses are present even when the system is inoperative. Nevertheless, operational time is the only reasonable parameter for logical treatment.

Reliability can be measured by the chances of failure at any point in the lifetime of a device or system. It is possible to chart this. The vertical axis will be probability of failure (one chance in six, one chance in three, etc.) and the horizontal axis will be the operating lifetime. It should be noted that workmanship failures occur first. This is why a manufacturer can honestly warrant a system. Beyond a certain time, wearout is the predominant cause of failure.

## WRITING AND INTERPRETING SPECIFICATIONS

There are two approaches to preparing the specifications when contractural procurement procedure is followed. The first method is to shop around among the various manufacturers and prepare specifications from those submitted by selected manufacturers. The second approach is to submit objective specifications based upon requirements and hope that they will be met by bidders. Many times a combination of the two approaches is used. The latter approach would seem the better one, but also the more difficult to pursue.

The specifications writer may or may not be familiar with the state of the art. In any case, the first step will be to gather as much information as possible about available equipment and costs. This is often not a simple task. Equipment manufacturers do not always have adequate information available in a concisely written form. In general, the dealer or manufacturer who can provide concrete information regarding performance will be the one who can do the job. It is also of help to visit existing installations to interview personnel actually using the systems.

Video surveillance equipment may be either "off-the-shelf" or customized for a specific task. Existing models with a history of reliability cost less and are safer to specify than customized equipment. Special designs will require that the entire engineering cost be written off on a single order. This could mean equipment that is inadequately designed and tested.

The idea in writing objective specifications is to specify the end performance in a practical situation and let the supplier agree to meet it. For example, an objective specification would be:

The system will be capable of distinguishing the features of an adult subject at a distance of 25 feet from the lens, at an average light level of 0.5 foot-lambert.

The other approach would be to specify lens speed, resolution, type of vidicon, etc. This narrows the procurement choice. As a matter of fact, most manufacturers prepare specifications that are handed out to architects and engineers. This practice narrows the choice to the manufacturer's products, thus "locking in" the bid. The practice in itself is not necessarily bad, but the problem is that the manufacturer's specifications may occasionally be poorly written and in fact do not specify actual parameter limits. We can look at the specifications of a video surveillance system under the following three headings:

1. Performance (ability to do the job).
2. Environmental Survival (ability to perform in existing environment).
3. Reliability and Wear (electronic mean-time-to-failure and mechanical wear).

A good set of specifications should encompass these three factors for the entire system. After the configuration of a system has been decided upon, each element of the system should be examined from the standpoint of performance, environment, and reliability. For a solid-state electronic component used indoors, the reliability and environment specifications will probably not require a great deal of attention, but electrical performance will be a large concern. A pan and tilt drive on the other hand will require both environmental and reliability considerations. In specifying environmental requirements, unique situations such as the presence of corrosive atmosphere, extreme heat, etc., must be considered.

Finally, it should be remembered that contractual specifications should protect and enhance the final installation. Poorly thought out specifications can do just the opposite. It should be specified that the manufacturer provide reasonable warranty and that instruction manuals and schematics be supplied. In some cases, the presence of a factory technical representative may be required.

# 13

# *Installations*

Video installations for surveillance purposes may cost from a few hundred dollars to hundreds of thousands. A video installation may be used for security purposes, in parking lots, retail outlets, etc.; or it may be intended for monitoring, as in hospital patient care or industrial control. In any case, the installation should be based upon economic parameters. Either the cost to be committed is based upon the resultant savings, or the use of CCTV should result in economic gain by doing something that otherwise cannot be done. Up until now the discussion has centered around individual components. Now we may examine the final problem—actual installation.

## PLANNING A SYSTEM

By the time the planning stage has been reached, the decision has usually already been made to commit funds for a system. A rough cost may also have been arrived at. The planner still may have considerable latitude, however.

A surveillance system for security purposes is for the conveyal of physical presence and activity. This may be done by proximity or intrusion alarms, by video means, or by a combination of these. In any case, the system is comprised of two elements: the camera or alarm points and the monitoring centers. At this point in the text, the reader has become familiar with the simple one or two-camera hookups. We can now proceed to a more sophisticated example of an installation in a large discount retail store.

Our store will carry a wide merchandise selection in three categories: wearables for everyone; sporting goods, appliances, records, etc.; and home building items. Each of the three categories is divided and physically separated with separate checkout facilities. In addition, there is a large adjoining warehouse. The video system must contribute to surveillance of the warehouse where only employees have access, and must guard against shoplifting in customer areas. Three security personnel will be on duty during opening hours, a male and female operative and one uniformed guard.

The solution can begin by locating the alarm and monitoring center so that cable runs are minimum. If the installation is made after the building is constructed, it may be advantageous to use 24 VAC cameras and controls in order to lessen the expense of wiring. Low voltage systems are, in general, less efficient, and the choice of equipment is limited, however. One security person will be present at the alarm center at all times. The alarm center will communicate with security persons on the floor by overriding the public announcement system using messages disguised as announcements.

With reference to the floor plan of Figure 13-1, we decide to locate the alarm and monitoring center in the specially constructed security office. Entrance to the security office is through unobtrusively located doors. Twelve cameras comprise the video set-up, located at A, B, C, . . . etc.

The most elaborate cameras are A, B, C, D, E and F. These are ceiling mounted in discreet domed housings with one-way vision and equipped with pan and tilt drives with auto–pan. A further feature is that each camera is controlled by a preset position unit and equipped with zoom lens. Thus, the operator has a couple of choices of maneuvering the camera: he may manually move the camera in pan and tilt and zoom into an area; or he may position the camera in tilt, set the zoom, and cause the

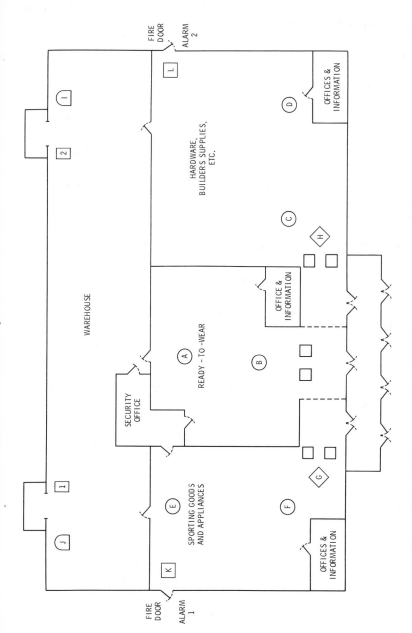

**Figure 13-1.** Example of a Video Surveillance System in a Retail Discount Store.

camera to drive to one of four areas and zoom in. These areas have been chosen as potential areas of interest. The present positions also help the operator in orienting himself prior to manual maneuvering.

Cameras G and H are fixed, equipped with fixed focal length wide angle auto-iris lenses, and cover checkout areas. Cameras K and L are equipped with fixed focal length lenses and cover the fire doors. Alarms 1 and 2 automatically place these cameras on a monitor screen when the door is opened. Otherwise, video output from these cameras is disconnected. Cameras I and J are fixed, with fixed focal length auto-iris lenses. They cover warehouse doors.

This setup covers daytime operations. It also covers night surveillance in the event that this is found necessary. For night surveillance, it may be adequate to shut down cameras A, B, C, D, E, and F. On the other hand, outdoor surveillance cameras covering the surrounding area may be added for night use, allowing a single guard to monitor from the security office. The store may also be protected at night with an unmanned alarm system.

It should be noted that this is a general example of a video surveillance application. Regional laws and building requirements will have an effect on an installation, as well as operating policies. This is an elaborate system. One security operative is on duty at the monitoring center at all times. With this system, it is possible to observe a theft, and to track and apprehend a perpetrator before he has left the store. The actual procedure will depend upon store policy, discretion, and whether or not security personnel have power of arrest status.

In our example, the monitoring center is concealed from both the public and employees. Some installations, however, place the monitoring center on public view for psychological purposes. Monitors are usually arranged in customized consoles. Consoles are available on a building-block basis. Video equipment is built for standard 1 3/4" X 19" EIA racking. Controls may sometimes be mounted loose on rubber feet for desk-top use.

Figure 13-2 shows the arrangement of the monitoring center. Three monitors are used. The two monitors on the right are for the elaborate cameras in each of the three store sections. A bridging sequential switcher displays the output of each camera in adjustable time sequence on the right monitor. When the operator wishes to observe an individual camera, he presses one of six switches and the camera output appears on the left most monitor.

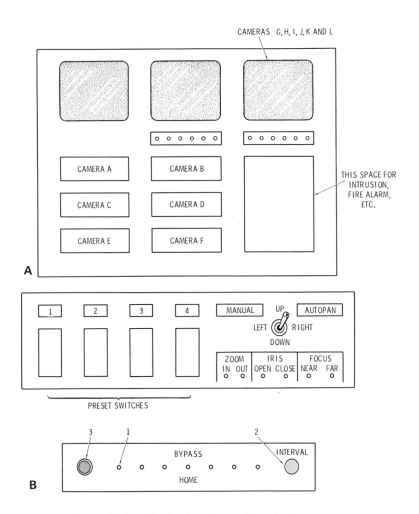

CAMERAS  G, H, I, J, K AND L

CAMERA A          CAMERA B

CAMERA C          CAMERA D

CAMERA E          CAMERA F

THIS SPACE FOR
INTRUSION,
FIRE ALARM,
ETC.

A

1        2        3        4        MANUAL      UP      AUTOPAN

LEFT      RIGHT

DOWN

ZOOM        IRIS        FOCUS
IN  OUT  OPEN CLOSE  NEAR  FAR

PRESET SWITCHES

3        1                              2

BYPASS              INTERVAL

HOME

B

**Figure 13-2.** Monitoring Center Console Arrangement.

The control for each camera is shown in Figure 13-2B. When it is desired to go to a preset position one of the respective buttons is depressed. When manual control of a camera is desired, the MANUAL switch is depressed. The controls on the right may then be used. For auto-pan operation, the AUTO-PAN button is pressed. In this position, focus, zoom, iris, and tilt up-down may be manually controlled, but pan is continuous back and forth.

The fixed cameras are controlled from the switcher under the right-most monitor. Only the output of cameras G, H, I, and J appear in sequence. When, however, any of the two fire doors are alarmed (opened), the respective camera output automatically appears. Operation of the sequential switcher is simple. See Figure 13-2C. A row of buttons (1) controls each camera. Placing a switch in the up (BYPASS) position takes the camera out of the sequence. The down (HOME) position stops the sequence and retains the respective camera on the monitor. When a fire door is alarmed, the annunciator (3) sounds and the respective camera output is displayed.

From this example it can be seen that a video surveillance system can be made very effective. After a short while operators become very adept at interpreting the monitors and operating cameras. To be effective, the video system should avoid gimmickry and be well planned. If, after a new system has operated for a while, it is found that it is not providing the expected coverage, it may be necessary to reevaluate the requirement and make changes.


## ILLUMINATION AND CAMERA PERFORMANCE

When the location of a camera has been determined, its performance may be evaluated by determining the amount of light reaching the faceplate. Theoretically, this can be done before cameras are installed or even purchased, as will be shown.

The amount of light reaching the lens depends upon illumination intensity and reflectance. A scene may be illuminated by sunlight, starlight, flood light, etc. The amount of light gathered by the lens for a given amount of illumination depends upon the reflectance of objects within the field of view. For example, white objects have the highest reflectance.

One way to determine the requirements for effective surveillance is to actually set up a camera under worst case conditions of brightest and dimest scene illumination. This will only reveal the capabilities of a specific camera, however, and will not necessarily specify the performance required for the situation.

Figure 13-3 shows the principle of scene illumination. In the figure, A is a point source of illumination. Light from this source is then reflected through the diaphram (B) and the lens (C).

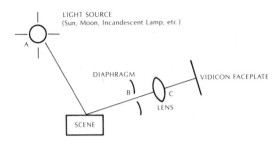

**Figure 13-3.**  Scene Illumination.

In most practical situations, however, total illumination is not provided from a simple source, but arrives at an object from many directions. Skylight, especially at night near cities, is light which has no defined point of origin.

A camera is rated by the amount of light falling upon the vidicon faceplate. This will, in turn, be a function of the type of vidicon, its color response, and such factors as scan rate, amplifier efficiency, etc. These things are governed by the choice of camera. When the lens is installed, lens transmittance and the f-stop will control the amount of light which passes to the vidicon faceplate.

Light from the scene depends upon the *reflectance* of the scene. To determine the camera requirements for surveillance of a given scene, it is necessary to measure the light from the scene. This can be done with a suitable light meter which is calibrated into actual light units, rather than the standard units used in photography. The standard unit of measurement in scene lighting is the *foot candle.* It is a measure of the intensity of light falling on one square foot of surface, and is equal to one *lumen* per square foot (the lumen is a measure of light intensity). Another unit is the *lux*, or lumens per square meter. Still another is the *phot*, or lumens per square centimeter. These units may be converted as follows:

| To convert from | to | multiply by |
|---|---|---|
| Lux (lumens per square meter) | foot candles | 0.093 |
| Phots (lumens per square centimeter) | foot candles | 930.0 |

The problem is to choose a lens and camera combination which will satisfy the surveillance requirement when the illumination of the scene is measured with a light meter. This requires that light measured by the light meter must be converted to faceplate illumination, and some method must be found to reconcile the measured illumination with the specifications given by the camera manufacturer. Cameras are usually rated in some manner related to minimum light level for usable picture. Here are some representative examples of camera ratings which actually can apply to the same camera and lens:

- $2 \times 10^{-6}$ foot candle faceplate illumination for useable picture
- Starlight illumination of $10^{-4}$ foot candles with 0.2 average scene reflection and f/1.5 lens.
- $10^{-4}$ scene illumination with f/1.5 lens

These figures all describe a camera of identical performance, but certain factors are left out of each rating. A list of the factors relating to surveillance requirements would be:

- lens f-stop (f)
- lens transmittance (T)
- scene reflectance (R)
- pickup tube sensitivity (S)

Scene illumination will be as shown in Table 13-1. Scene reflectance depends upon the characteristics of objects within the scene. Transmittance deals with light lost while passing through the lens. This table can be used to reconcile the various ways of specifying performance.

The actual performance, however, is the amount of faceplate illumination required to reproduce scene details. In practical cases, the light meter measurement (I) will be:

$$I = SR$$

Lens transmittance (T) may not be known, but since the f-stop can be changed to suit a given scene, this can be compensated for by changing the f-stop with the lens iris. In practical cases,

**Table 13-1.**   Outdoor Scene Illumination*

| | | |
|---|---|---|
| Direct sunlight | 10,000 | $(10^4)$ |
| Full daylight | 1,000 | $(10^3)$ |
| Overcast day | 100 | $(10^2)$ |
| Very dark day | 10 | $(10^1)$ |
| Twilight | 1 | $(10^0)$ |
| Deep twilight | 0.1 | $(10^{-1})$ |
| Full moon | 0.01 | $(10^{-2})$ |
| Quarter moon | 0.001 | $(10^{-3})$ |
| Starlight | 0.0001 | $(10^{-4})$ |
| Overcast night | 0.00001 | $(10^{-5})$ |

*This shows the extreme range of light variation from an overcast night ($10^{-5}$ foot candles) to direct sunlight ($10^4$ foot candles). This is a variation of one million to one. A video pickup tube capable of operating at one end of this range could not operate at the other without some arrangement to attentuate light.

it is possible to use a light meter value. The equation for this will be:

$$F = \frac{I}{4f^2}$$

where (I) is the light meter measurement. This assumes that the value of (f) is increased to allow for light lost in the lens. In most cases, an increase of one f-stop is sufficient.

An example of the method of calculation is the following situation: The interior of a store is found to have a minimum of 50 lumens reflected light from the scene as measured by a light meter. A certain camera will reproduce a satisfactory image using a f 2.8 zoom lens with 0.5 lumens faceplate illumination. The problem is to determine if this camera is adequate for surveillance of the store interior. Since the transmittance of the lens is not known, we allow one f-stop for light loss in the lens. The following is now known:

- Scene illumination and reflectance (this has been measured with a light meter).
- f-stop

It is necessary to find F, the faceplate illumination. Thus,

$$F = \frac{50}{4 \times 3.8 \times 3.8} = 0.87 \text{ ft-candles (lumens)}$$

If the camera will reproduce an adequate image with 0.50 foot-candles, this lens/camera combination will be adequate.

The foregoing method of analysis should always be used in the design of the larger, more costly installations. Camera specifications should be carefully interpreted. The important parameter is faceplate sensitivity. There will be a given video signal level for a given light level on the faceplate. This value alone, however, does not completely evaluate performance. Scan speed is also a factor. The faster the scan, the less useable the picture image at the lowest light levels will be. In other words, at the lowest light levels, the picture becomes more blurry. That means the apparent resolution and contrast is less. Figure 13-4 is a chart showing the performance of a camera capable of operating at very low light levels. At the lowest practical faceplate illumination, resolution drops to the equivalent of less than 150 TV lines. At light levels above one-tenth of a foot-candle, the video signal normalizes at 700 TV lines, where it remains for larger light values.

Another performance criteria is the sensitivity of the camera tube to the light spectrum. Certain pickup tubes will be more responsive to the infrared spectrum. Artificial lighting of the scene has more components toward the yellow spectrum, and so on. The spectrum response of vidicons is available from the manufacturer. In some cases it may be necessary to actually test the camera rather than resorting to light meter measurements which do not necessarily reveal spectrum characteristics. An example would be an entertainment area that is deliberately illuminated with colored lights, or even changing colors.

## LABOR, WIRING AND CABLE COSTS

The location of monitoring centers is dictated by various factors, one being wiring and cable costs. Obviously, every effort will be made to keep the distance as short as possible. Another factor which cuts wiring costs is the choice of protected areas for coaxial runs which permit the use of cheaper indoor type cable. A certain

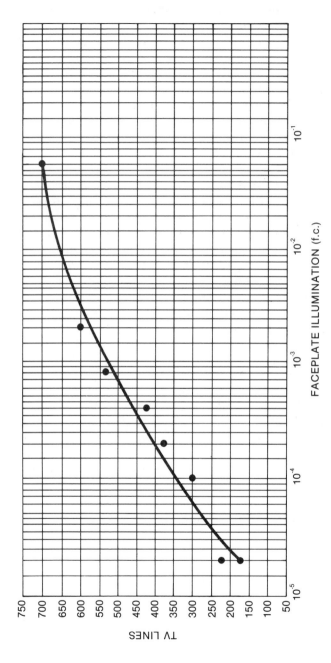

**Figure13-4.** Performance of Camera Capable of Operating at Low Light Levels.

amount of flexiblity may be necessary in cases where situations change such as in warehouses, etc.

A great deal of use is often made of video connectors. In some cases this may be more cosmetic than practical, although when equipment is occasionally shifted or withdrawn for maintenance and repair, connectors are advantageous. In any case, units that are "hard wired" are less costly and installation costs are usually less. It is important, however, to consider the matter of system flexibility. Warehousing situations change and retail customer facilities usually change as often as once a year. The same applies to industrial facilities.

Costs are estimated on a labor and materials basis. Individual units may be off-the-shelf or custom items supplied upon bid or catalog selection. Wire and cable runs are costed out as interior or exterior. Normally there would be two classes of labor involved: licensed electrical contractor or video technician. The video technician installs coaxial cable and low voltage wire. The licensed electrician installs wiring that is affected by building the electrical codes. In most areas, a video technician can make a complete 24 volt installation or any installation that received primary power from existing convenience outlets.

Video controls which are actuated by supplying power over wires originating at the controls are rapidly becoming outmoded in favor of methods employing signalling, with power coming from nearby convenience outlets. This method is much less costly to install and at least equally as reliable. In a permanent installation, especially where tamper proofing is important, electrical conduit is provided for incoming power and connection made through fittings.

When estimating cable lengths, 3000 feet is the most distance that a standard camera signal can usually be sent without amplification (or "equalization"). The following applies:

Attenuation per 100 feet at 8 MHz.

|         |          |
|---------|----------|
| RG–59U  | 1 db.    |
| RG–11U  | .55 db.  |
| RG–11U  | .35 db.  |

When using 24 volts to power cameras, the following can be used for estimating:

| Wire size, AWG | Maximum distance |
|:---:|:---:|
| 18 | 100 ft |
| 16 | 300 ft |
| 14 | 500 ft |
| 12 | 700 ft |

## OUTDOOR CABLE INSTALLATION

For long outdoor runs, when calculating cable attenuation, use the attenuation at the highest frequency of the usable bandwidth (see Figure 13-5). For example, when working with a 6-MHz bandwidth, use cable attenuation at 6 MHz. The same procedure applies when specifying equipment, for example "25 db of

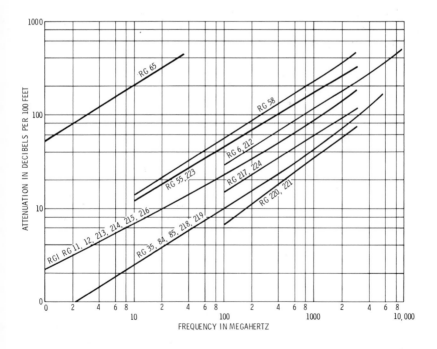

Figure 13-5.   Cable Attenuation.

equalization at 6 MHz. " Typical cable attenuation curves are such that an equalizer which will provide up to 30 db of equalization of 10 MHz would provide only approximately 15 db of equalization at 6 MHz. Thus, frequency should always be specified.

Avoid splicing on long cable runs, and do not use connectors unless absolutely necessary. Specify a single length of cable for the entire run where possible. Outdoor and underground connections must be impervious to moisture. Avoid running cable adjacent to highpower RF sources, power lines, or in the area of heavy electrical equipment or other hum–producing sources. A good earth ground is essential with long transmission lines. Poor grounding will result in a ground potential difference between each end of a cable run, introducing hum on an unbalanced line. Ground the equipment to a master ground connected to a water pipe located in the vicinity of the equipment.

Temperature variation should be taken into consideration. In temperate climates, variation between −5 degrees F and +110 degrees F are possible. Line–equalization equipment must be capable of acting throughout the attenuation range at these temperatures.

On complex systems, initial planning should include a diagram showing length and attenuation in db of each cable run between equipment, power sources, types of cable, methods of cable, and equipment installation such as direct earth burial, pole-mounting, etc. as well as signal source and monitoring equipment. Where cables cross casements, streets, or private property, rights of way must be procured.

When installing cable, always ground long cables at the connector before mating them to the equipment. Excessive current may appear at the end of a long cable which can damage equipment. The voltage may also be of sufficient potential to be harmful.

## LIGHTNING PROTECTION

Lightning protection is often overlooked in video security installations. Any exterior conductor is subject to lightning activity. Unless protection is provided, the result is always damaging to some degree, often catastrophic. Security installations that include exterior wire runs, either in the form of coaxial cable or control and power wires, should be protected against lightning damage.

The economics of lightning protection require that the incidence of lightning storms in each local area be known. The U.S. Weather Bureau can provide maps showing the susceptability of various locales to lightning. The probabilities of a lightning strike in a given area can be used as a factor to determine the attention given to lightning protection.

Lightning is the sudden and violent discharge of electrons from the clouds to the ground or between clouds. Hence, clouds are negatively charged bodies. Thunder is the sound of millions of volts arcing between clouds.

Studies show that a lightning stroke usually lasts about three tenths of a second, although as much as one and a half seconds have been recorded. A stroke is actually a series of arcs beginning with a "pilot leader," followed by other leaders moving from the cloud to the ground at between 100 and 2600 centimeters in a millionth of a second. When the leaders reach the ground a main streamer extends from the ground to the cloud at up to one half the speed of light. There may be as many as 30 discharges of this kind in each lightning stroke. The stroke usually extends from 2000 to 7000 feet up, although heights of 16,000 feet have been recorded.

In general, lightning strikes an object when it is closer to ground potential than other immediate objects. The higher the object in relation to others, the more likely it is to be struck. There is no way to predict the path the strike will take, however. Lightning has been known to choose a path along the ground, then climb an object and leap to another.

Direct lightning strike is violently destructive, and in most cases it is agreed that electronic equipment cannot be protected from direct strikes. On the other hand, most lightning damage is caused by coincidental potential differences which do not involve a direct strike, but result from surrounding activity. In a lightning–prone atmosphere, there is constant electron movement to elements which are in ground contact. A charge may be induced by adjacent activity. Damage from this can be prevented.

The telephone industry has involved itself in lightning protection since the first telephone lines were strung over a century ago. Devices developed by the telephone industry are available to the security system installer. They are listed in telephone supply catalogs.

Protective devices operate on the principle of charge ionization. If two electrodes are separated by air or certain gaseous matter, when the potential between the electrodes becomes high enough, ionization occurs between the electrodes. The ionized zone becomes a low resistance path and an arc occurs. When the potential is removed, the ionized path is removed. This is the principle used in lightning protection devices. In the absence of a high potential there is no conductive path between the electrodes, and the associated conductor may carry on a normal function in a system. One electrode is attached to the ground, the other to the conductor to be protected. This simple concept is illustrated in Figure 13-6.

Protective devices take two general forms. The older type uses two carbon electrodes separated by air. Carbon supports an arc once it is formed when the air between becomes ionized. The arc has very low resistance, and conducts the charge directly to ground. The second type of device uses two electrodes in a gas-filled tube. The principle is the same, but ionization can take place at a lower potential, thus offering greater protection. Also, the arc forms in a consistently lower fraction of time. Since the high voltage surge is very rapid, this further guards the associated equipment from receiving a surge.

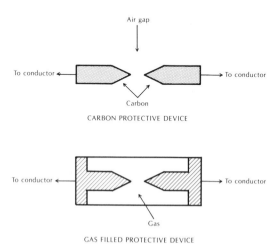

**Figure 13-6.** Devices for Protecting Against High Voltage Surge.

Another feature of the gas tube is longer life. Carbon electrodes deteriorate with each arc. Gas tubes recover and act virtually the same with each voltage surge.

Grounding is an important consideration in all lightning protection. A voltage surge will always follow the path of least impedance. Figure 13-7A shows two paths to ground. It can be assumed that both are of heavy wire and approximately of equal resistance. The surge will follow the path which has the least kinks and turns. The reason for this is that a turn represents an impedance to the rapidly rising voltage surge. Hence, use the straightest path to ground.

Exterior overhead runs of coaxial cable are susceptible to surge because they are nearer to ground potential than many adjacent objects. Figure 13-7B shows how electronic equipment can be protected from surges from the cable shield. A straight ground wire is interposed on the shield before the cable enters equipment.

A grounding point must offer the least possible resistance to ground. Water pipes are considered good. Concrete does not offer a good ground. Ground rods may be used. To be effective, a ground rod must be driven several feet down.

In conclusion, it is recommended that the video security installer make use of telephone practice in providing lightning protection. Publications and specifications on this subject are available.

## EQUIPMENT INSTALLATION

After new equipment is received from the manufacturer, the first step is a bench check. The bench check is important for two reasons: first, it will eliminate problems at the onset under circumstances that make corrections and repairs possible at less cost; second, certain preliminary adjustments can only be conveniently made on the bench.

It is important for the manufacturer to furnish complete adjustment information and schematics. Preliminary adjustments vary with the various models. Certain shop equipment must be available such as a good triggered oscilloscope with calibrated time base. A good monitor which has been previously tested should be handy. Much time can often be saved by substituting and exchanging units to locate problem areas before attempting repairs.

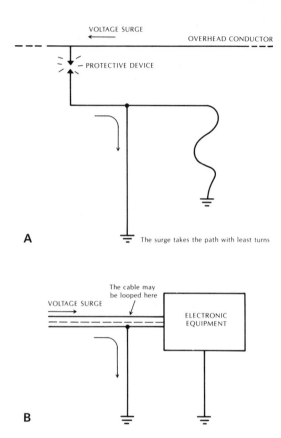

**Figure 13-7.** Grounding Methods for Lightning Surge Protection.

When separate external synchronization is to be employed, set up the entire system on the bench first and use the sync mode to be used in the final installation. Follow the manufacturer's instructions and make sync level and setup adjustment at this time. The test setup should include any switchers to be used. When a switcher appears faulty, first check camera and monitors by connecting them directly. Check the connectors. Make the bench check as thorough as possible so that, when the final hookup is made, any system faults can be traced to the transmission system (cables and equalizers or line amplifiers) only.

Camera–lens–monitor combinations should first be set up and operated on the bench. Units may be adjusted for focus at this time. Focus depends upon several interrelated factors. When a new system is received it is likely that the various elements will require focusing. The following remarks apply:

- *Electronic Focus*: This applies to the focusing of electron beams in camera and monitor.
- *Mechanical Focus*: The vidicon in the camera is usually provided with a means of moving the image surface with relation to the lens.
- *Optical Focus*: A focus adjustment is provided on each lens. Once the optical focus is properly set, elements of a zoom lens are driven in relationship to each other so that a sharp image is projected on the vidicon for all zoom settings.

When making the bench focus check, use short test cables to eliminate picture deterioration due to frequency component attenuation. Position the camera about 15 feet from a suitable focusing target or textured surface. Begin by focusing the electron beams, then proceed to optical and mechanical focus. Focusing should be under conditions of subdued lighting so that the iris may be opened. Under these conditions, the depth of field will be the least and small focusing errors will produce a more noticably fuzzy image. Repeat the order of adjustment until the sharpest image is obtained. A specific procedure for motorized zoom lens focusing is as follows:

1. Press IRIS OPEN button on control to open iris fully.
2. Operate ZOOM IN control to set lens at extreme telephoto position (smallest field of view) and focus on target using the FOCUS button.
3. Check electronic focus on camera and adjust if necessary.
4. Without changing focus, zoom OUT slowly toward the wide angle position (largest field of view). While zooming, check focus.
   a. If lens goes completely out of focus, adjust the picture at that point by use of mechanical focus adjustment on camera.

*Note*: This applies only to cameras with provisions for mechanical position focus. If focusing difficulty is encountered with factory preset mechanical focus, the position of the sensitive surface must be recalibrated to the .690" dimension from the lens "C" mount backing surface.

b. If the lens retains partially satisfactory focus, continue zooming to the OUT position and adjust the mechanical focus for best focus at that point.

5. Zoom to the extreme IN position and refocus the lens.
6. Repeat steps 3 and 4, fine trimming mechanical focus each time until the lens holds focus throughout its range.
7. As a final check, readjust electronic focus for optimum definition.

## TEST PATTERNS FOR EVALUATING SYSTEM PERFORMANCE

Several test patterns are available for televising to evaluate a system. A test pattern may be any image with characteristics that generate video signals of critical waveshapes. A special pattern may be reproduced by printing it on cardboard, or it may be on transparent acetate or mylar for use in a light box. The light box is a superior method since it allows uniform faceplate illumination and light characteristics can be controlled.

A standard test pattern is shown in Figure 13-8. It acts as a standard reference for measuring resolution, streaking, ringing, interlace, shading, scanning linearity, aspect ratio, and grey scale reproduction. The camera is set up so that the test pattern covers the entire viewing field. Horizontal resolution of a system may be limited by the camera tube, bandwidth of the video amplifiers, or interconnecting cable. Information concerning resolution, percent response at various line numbers, and degradation of resolution with camera tube aging can be obtained from a test chart containing a high number of lines. For these reasons, the horizontal and vertical wedges of the resolution chart are arranged to permit resolution measurements from 200 to 800 lines for the standard chart and from 200 to 1600 for a high-resolution chart.

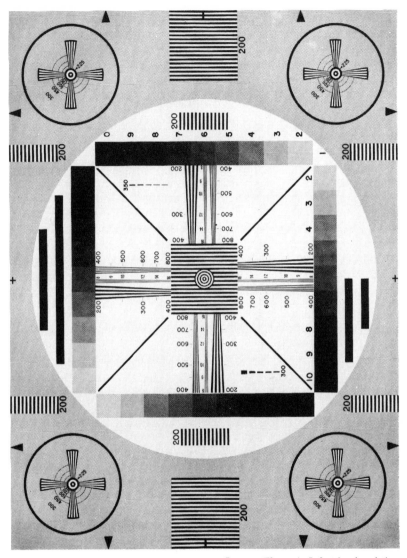

Courtesy Electronic Industries Association

Figure 13-8.  Video Test Pattern.

The center horizontal and vertical wedges are composed of four black lines separated by three equal-width white lines. Numbers printed alongside the wedges correspond to the total number of lines of the indicated thickness that may be placed adjacent to one another in the height of the chart. For example, if black and white lines having the same thickness as those indicated at the 320 position were placed adjacent to one another, a total of 320 lines could be fitted into the height of the chart. Since the aspect ratio of the chart is 4 to 3, a total of (320 × 4/3) or 426.7 of these same thickness lines could be placed in the width of the chart. The fundamental video frequencies generated in scanning through various parts of the vertical wedges are tabulated in Table 13-2.

Shading may be checked on the monitor by noting if the background is an even gray and if the same number of gray steps are discernible on all four gray scales. A waveform monitor may also be used to determine if the average picture axis is parallel to the blanking-level line at both line and field frequencies. Streaking of the horizontal black bars at the top or bottom of the large circle is an indication of low-frequency phase shift or poor dc restoration. The black bars are also used for adjusting the high-peaking circuits used to compensate for the high frequency roll off. The four diagonal black lines inside the square formed by the gray scales may be used to check interlace. A jagged line indicates pairing of the interlaced lines.

**Table 13-2.** Fundamental Video Frequencies Generated in Scanning Through Various Parts of Vertical Wedges (EIA television standards)

| Line number on vertical wedge | Fundamental video frequency | Line number on vertical wedge | Fundamental video frequency |
|---|---|---|---|
| 200 | 2.5 MHz | 880 | 11.0 MHz |
| 240 | 3.0 MHz | 960 | 12.0 MHz |
| 280 | 3.5 MHz | 1040 | 13.0 MHz |
| 320 | 4.0 MHz | 1120 | 14.0 MHz |
| 400 | 5.0 MHz | 1200 | 15.0 MHz |
| 480 | 6.0 MHz | 1280 | 16.0 MHz |
| 560 | 7.0 MHz | 1360 | 17.0 MHz |
| 640 | 8.0 MHz | 1440 | 18.0 MHz |
| 720 | 9.0 MHz | 1520 | 19.0 MHz |
| 800 | 10.0 MHz | 1600 | 20.0 MHz |

The two sections of single line widths at the upper right hand portion and lower left hand portion of the square formed by the gray scale may be used to check ringing. These lines are included because the multiple lines in the wedges are confusing. Lines in the upper right hand section have widths from 350–550 (350, 400, 450, 500, 550), and lines in the lower left hand section have widths from 100–300 (100, 150, 200, 250, 300).

Another useful test pattern is a white square window on a black field with a 4:3 aspect ratio. Before the test, the camera and monitor must be adjusted for scan size and linearity. Adjust the camera operating controls for normal operation. Observe the video signal on an oscilloscope. A square wave, resulting from abrupt changes from black–to–white–to–black, should be seen. A slope to the top of the square wave indicates a phase shift at the low frequencies. If the top of the square wave is concave or convex, this indicates amplitude distortion at the low frequencies. Low frequency smear, phase shift, or misadjusted high peaker is indicated by trailing white or black streaking where abrupt changes occur from black–to–white or white–to–black.

# Glossary

With every new industry comes special terminology and jargon. In some cases a term comes to mean different, even opposite, things. It is hoped that this glossary will help reveal the meaning of terms to new persons in the field and, secondly, serve to standardize terminology.

| | |
|---|---|
| *Absolute Maximum Rating.* | The operating limits of a device which, if exceeded, may result in permanent damage. |
| *ABC.* | Automatic Beam Control. |
| *AFC.* | Automatic Frequency Control. A circuit that automatically maintains the frequency of an oscillator within specified limits. |
| *AGC.* | Automatic Gain Control. A circuit for automatically controlling amplifier gain to maintain a constant output voltage with a varying input voltage. |
| *Alpha* ($\alpha$). | The current-amplification factor of a transistor when connected in a grounded-base configuration. The ratio of the incremental change in collector current to an incremental |

change in emitter current at a constant-collector potential.

*Amplifier.* A device for changing (increasing) the amplitude of a signal without altering its quality.

*Aperture.* The opening of a lens which controls the amount of light reaching the surface of the vidicon tube. The size of the aperture is controlled by the iris adjustment. By increasing the f stop number (f 1.4, f 1.8, f 1.9, f 2.8, etc.), less light is permitted to pass to the vidicon.

*Aspect Ratio.* The ratio of width to height of a television picture; four units wide by three units high in standard TV systems.

*Attenuation.* A loss in signal.

*Audio (AF).* Signals with a frequency which fall within the audible range, generally taken to be from 20 to 20,000 c.p.s.

*Auto-Block.* Circuit used to maintain a constant red or black level of camera.

*Auto-Pan.* Continuous, automatic horizontal back and forth motion of surveillance camera.

*Auto-Scan.* See Auto-Pan.

*AVC.* Automatic Volume Control.

*Back Porch.* That portion of the composite picture signal which lies between the trailing edge of the horizontal-sync pulse and the trailing edge of the corresponding blanking pulse.

*Bandpass.* A specific range of frequencies that will pass through a device.

*Bandpass Filter.* A filter that will pass only a specific band of frequencies.

*Bandwidth.* The number expressing the difference between the lower and upper limiting frequencies of a device. It is usually taken between points at which the power is one half of maximum (3 db.).

| | |
|---|---|
| *Barrel Distortion.* | Distortion that makes the televised image to appear to bulge like a barrel. |
| *Base.* | One of the electrodes in a transistor. Roughly analogus to the grid or "control electrode" of a vacuum tube. |
| *Beta (β).* | The current amplification of a transistor when connected in the grounded–emitter configuration. Essentially, the ratio of the incremental change in collector current to an incremental change in base current at a constant–collector voltage. |
| *Bias.* | The steady (DC) operating voltage or current applied to an electrode to establish the basic operating conditions of a device. A bias current establishes the operating conditions of a transistor. |
| *Black Level.* | The level of a television signal that corresponds to the maximum limits of the black of the picture. |
| *Blanking.* | The process of cutting off the electron beam in a camera or picture tube during the retrace period. |
| *Blanking Level.* | The level of a television signal which separates the range containing picture information from the range containing synchronizing information; also called pedestal. |
| *Blanking Signal.* | A signal related in time to the scanning process used to effect blanking. |
| *Blooming.* | Halation and defocusing around bright picture areas on monitor screen. |
| *Breezeway.* | That portion of the back porch between the trailing edge of the horizontal synchronizing pulse and the start of the color burst. |
| *Bridging.* | A term indicating that a high impedance video line is paralleled, usually through a switch, to a source of video. |
| *Brightness.* | A monitor adjustment which varies the overall brightness of the televised picture. |

*Cathode-Ray Tube.*   The picture tube in a video monitor which reproduces the picture image seen by the camera.

*Candlepower.*   A unit measure of *incident* light.

*CCTV.*   Common abbreviation for Closed Circuit Television.

*Characteristics.*   The electrical specifications of a device, sometimes given as a table or curve showing the variation of one electrical quantity with respect to another.

*Chrominance Signal.*   That portion of the color television signal which contains the color information.

*Clamper.*   A circuit to hold the level of the picture signal at some predetermined reference level at the beginning of each scanning line.

*Clipping.*   Shearing off of the peaks of either the positive (white) or negative (black) peaks.

*Coaxial Cable.*   A type of cable capable of passing a range of frequencies with low loss. It consists of a hollow metallic shield with a single wire placed along the center of the shield and isolated from the shield.

*Collector.*   One of the electrodes in a transistor. Roughly analogous to the plate of a vacuum tube.

*Color Burst.*   The portion of a composite color signal comprising a few cycles of a sine wave of chrominance subcarrier frequency used to establish a reference for demodulating the chrominance signal.

*Complementary Symmetry.*   The term applied to circuits utilizing the similar (symmetrical) but opposite (complementary) characteristics of PNP and NPN junction transistors.

*Composite Video Signal.*   The combined picture signal, including vertical and horizontal blanking and synchronizing signals.

*Compression.*   The reduction in gain at one level of a picture signal with respect to the gain at another level of the same signal.

| | |
|---|---|
| *Configuration.* | The circuit arrangement. There are three basic configurations used in transistor circuitry ... (a) grounded-base, (b) grounded-emitter, and (c) grounded-collector. |
| *Contrast.* | The brightness difference within a televised subject. Low contrast is shown as mainly shades of gray, while high contrast usually is distinct blacks and whites with very little gray. Also a monitor adjustment which increases or decreases the level of contrast of the televised picture. |
| *Contrast Range.* | The ratio between the whitest and blackest portions of a television image. |
| *Crosstalk.* | An undersired signal interfering with the desired signal. |
| *Cutoff Frequency.* | Generally taken as the frequency at which the gain of a device is 3 db below its low frequency value. Used when referring to the variation of alpha or beta with respect to frequency. |
| *DB (Decibel).* | A measure of the power ratio of two signals. It is equal to ten times the logarithm of the ratio of the two signals. |
| *DC Restoration.* | The reestablishment by a sampling process of the DC and the low-frequency components of a video signal. |
| *Delay Distortion.* | Distortion resulting from the nonuniform speed of transmission of the various frequency components of a signal; the various frequency components of the signal have different times of travel (delay) between the input and the output of a circuit. |
| *Depth of Field.* | The front to back zone in a field of view which is in focus in the televised scene. With a greater depth of field, more of the scene, near to far, is in focus. Increasing the f-stop number increases the depth of field of a TV lens. Therefore, the lens aperture should be set at the highest f-stop |

number usable with the available lighting. The better the lighting, the greater the depth of field possible.

*Detector.* An electrical circuit or device used to remove the modulation from a carrier signal.

*Diode.* An electrical device having two electrodes—cathode and anode. May be either a semiconductor or a vacuum tube.

*Dissipation.* A loss of energy. Generally used to indicate the amount of electrical energy converted into heat by a device. Expressed in watts.

*Distortion.* Deviation of the received waveform from that of the original transmitted waveform.

*ECHO.* A signal which has been reflected at one or more points during transmission with sufficient magnitude and time difference as to be detected as a signal distinct from that of the primary signal. Echoes can be either leading or lagging the primary signal and appear as reflection, or "ghosts."

*Efficiency.* The ratio of useful power output to power input. Generally expressed as a percentage. A "perfect" device would be 100% efficient. Since the difference between power output and power input to a device represents a loss of energy, the more efficient a device, the lower the internal losses. A "perfect" device would have zero losses.

*Emitter.* One of the electrodes in a transistor. Roughly analogous to the cathode of a vacuum tube. An emitter junction is generally biased in a forward direction.

*Equalization.* The process of correcting losses of a signal. Usually applied at the receiving terminal.

*Equalizer.* Equipment designed to compensate for loss and delay frequency effects within a television system.

| | |
|---|---|
| *Feedback.* | In a mechanical or electrical system, returning a control signal from one part of the system to an earlier stage. The signal is "fed back," hence the term. |
| *Feedback, Degenerative.* | Also termed negative or inverse feedback. A feedback signal that is out-of-phase with the forward moving signal at the point at which it is injected into the system, thus tending to cancel a portion of the normal signal. |
| *Feedback, Regenerative.* | Also termed positive feedback. A feedback signal that is in-phase with the forward moving signal at the point at which it is injected into the system, thus tending to increase the amplitude of that signal. Regenerative feedback thus tends to continue changes in the signal in the same direction, increasing the system's gain. |
| *Field.* | One of the two equal parts into which a television frame is divided in an interlaced system of scanning. |
| *Field Frequency.* | The number of fields transmitted per second in a television system. The U.S. standard is 60 fields per second. Also called field-repetition rate. |
| *Field of View.* | The area that can be seen by the camera. |
| *Fly Back.* | The rapid return of the electron beam in the direction opposite to that used for scanning. |
| *Focal Length.* | The distance from the center of the lens to a plane at which point a sharp image will be produced of an object at an infinite distance from the camera. The focal length determines the size of the image and the angle of the field of view seen by the camera through the lens. Lenses are grouped by focal length expressed in millimeters (mm). |

*Frame.*  The total picture area which is scanned while the picture signal is not blanked.

*Frequency Response.*  The range or band of frequencies to which a unit of electronic equipment will offer essentially the same characteristics. See "Bandwidth."

*Front Porch.*  The portion of the composite picture signal which lies between the leading edge of the horizontal blanking pulse and the leading edge of the corresponding synchronizing pulse.

*Gain.*  The amount of amplification which a system gives to a signal. If the system reduces (attenuates) the signal, the gain is said to be less than unity (one). In electrical systems, voltage and current gains are often expressed in decibels.

*Gamma.*  A measure of contrast or gray scale in the television picture.

*Geometric Distortion.*  An aberration which causes the reproduced picture to be geometrically dissimilar to the perspective plane projection of the original scene.

*Ghost.*  A shadowy or weak image in the received picture, offset either to the right or to the left of the primary image.

*Glitches.*  A form of interference appearing as a narrow horizontal bar either stationary or moving vertically through the picture.

*Ground.*  In circuit work, a common connection point, either a metal chassis, a common terminal, or a ground bus. If a point has zero voltage with respect to ground, it is said to be at "ground potential."

*Grounded-Base.*  Also common-base or GB. A transistor circuit configuration in which the base electrode is common to both the input and output circuits. The base need not be connected directly to circuit ground, however. A grounded-base transistor circuit is roughly analogous to a grounded-grid vacuum tube circuit.

*Grounded-Collector.* Also common-collector or GC. A transistor circuit configuration in which the collector electrode is common to both the input and output circuits. A grounded-collector transistor circuit is roughly analogous to a grounded-plate or cathode-follower vacuum tube circuit.

*Grounded-Emitter.* Also common-emitter or GE. A transistor circuit configuration in which the emitter electrode is common to both the input and output circuits. A grounded-emitter transistor circuit is roughly analogous to a grounded cathode vacuum tube circuit.

*Harmonic.* A signal having a frequency which is an integral multiple of the fundamental frequency from which it is derived or related.

*Heat Sink.* As applied to transistor work, a mass of metal or other good heat conductor which serves to quickly absorb and to dissipate quantities of heat energy. Used to prevent overheating of a transistor or semiconductor's electrodes.

*High-Frequency Distortion.* Distortion effects which occur at high frequency. Generally considered as any frequency above the 15.75-KHz line frequency.

*Highlights.* The maximum brightness of the picture, which occurs in regions of highest illumination.

*Homing.* The process of displaying one of a number of camera outputs on a given monitor.

*Horizontal (Hum) Bars.* Horizontal bars, alternately black and white, which extend over the entire picture. They may be stationary or may move up and down. It is usually caused by a 60 Hz interfering frequency or a harmonic frequency.

*Horizontal Blanking.*  The blanking signal at the end of each scanning line.

*Horizontal Resolution.*  The maximum number of individual picture elements that can be distinguished in a single horizontal scanning line.

*Horizontal Retrace.*  The return of the electron beam from the right to the left side of the raster after the scanning of one line while the screen is blanked or cut off.

*Hue.*  Corresponds to colors such as red, blue, etc. Black, gray and white do not have hue.

*Hum.*  Electrical disturbance at the power supply frequency or harmonics thereof.

*Impedance.*  The opposition which a circuit or component offers to the flow of electric current. It is expressed in ohms and is equal to the ratio of the effective value of the voltage applied to the circuit to the resulting current flow. In AC circuits, the impedance is a complex quantity including both resistance and reactance. In DC circuits, the impedance is purely resistive.

*Incident Light.*  Light falling directly on an object.

*Insertion Loss.*  The signal strength loss when a piece of equipment is inserted into a line.

*Interlaced Scanning.*  A scanning process for reducing image flicker in which the distance from center to center of successively scanned lines is two or more times the nominal line width, and in which the adjacent lines belong to different fields.

*Jitter.*  Small rapid variations in a waveform due to mechanical disturbances, changes in the characteristics of components, supply voltages, imperfect synchronizing signals, circuits, et cetera.

*Lag.*  Percent of initial value of signal output after illuminance is removed.

*Lambert.*     A unit measure of emitted or reflected light.

*Leading Edge.*     The major portion of the rise of a pulse, taken from 10 to 90 percent of total amplitude.

*Looping.*     A term indicating that a high–impedance device has been permanently connected in parallel to a source of video. This is usually done by entering and exiting through a coaxial connector.

*Loss.*     A reduction in signal level or strength, usually expressed in DB.

*Low-Frequency Distortion.*     Distortion effects which occur at low frequency. Generally considered as any frequency below the 15.75–KHz line frequency.

*Luminance Signal.*     That portion of the NTSC color television signal which contains the luminance or brightness information.

*Modulate.*     To change or vary some parameter such as varying the amplitude (amplitude modulation) or frequency (frequency modulation) of a signal. The circuit which modulates a signal is called a modulator.

*Monochrome.*     Having only one color. In television, black and white.

*Monochrome Signal.*     In monochrome television, a signal for controlling the brightness values in the picture. In color television, the signal which controls the birghtness of the picture, whether displayed in color or monochrome.

*Negative Image.*     An image produced by a video signal which is reversed 180 degrees in phase and in which white and black areas are reversed.

*Noise.*     Random spurts of electrical energy or interference. It produces a "salt–and–pepper" pattern over the televised picture.

*Noise Figure.*     A measure of the noise present in the output of a device. Expressed in DB.

*NTSC.*      National Television Systems Committee that worked with the FCC in formulating standards for the United States color television system.

*Overshoot.*      The initial transient response to a change in input which exceeds the steady-state response.

*Pairing.*      Overlapping of alternate scanning lines resulting in reduced vertical resolution.

*Pan.*      Movement of the camera in a horizontal direction.

*Passive.*      A non–powered element of a system.

*Peak–to–Peak.*      The amplitude difference between the most positive and the most negative excursions of a signal.

*Photodiode.*      A two–electrode light–sensitive device. In general, the resistance between the two electrodes varies inversely with the amount of light falling on the light–sensitive surface.

*Phototransistor.*      A light sensitive transistor. May have two or three electrodes, although a two electrode device is more properly called a photodiode.

*Pigeons.*      Noise on monitors as bursts at a slow rate of occurrence.

*Power.*      The rate at which electrical energy is fed to or taken from a device. Expressed in watts, milliwatts, or microwatts.

*Preamplifier.*      An amplifier to increase the output of a low-level source that the signal can be further processed above the circuit noise level.

*Pre-Emphasis.*      A change in the level of some frequency components of the signal at the input to a transmission system.

*Random Interlace.*      A simple scanning technique commonly used in CCTV systems in which there is no external control over the scanning process.

| | |
|---|---|
| *Raster.* | The rectangular pattern of scanning lines upon which the picture is produced. The illuminated face of the monitor. |
| *Reference Black Level.* | The signal level corresponding to a specified maximum limit for black peaks. |
| *Reference White Level.* | The signal level corresponding to a specified maximum limit for white peaks. |
| *Reflected Light.* | Scene brightness or light being reflected from a scene 5% to 95% of incident light measured in foot–lamberts. |
| *Repeater.* | Equipment used for receiving, amplifying, and retransmitting a signal on long transmission line. |
| *Resolution.* | A measure of the ability of a television system to reproduce detail. Resolution is limited by the lowest rated component (camera or monitor) in the TV system chain. |
| *Resolution (Horizontal).* | The amount of detail in the horizontal direction of a picture. It is expressed as a number of distinct vertical alternately black and white lines which can be seen in the picture height. |
| *Resolution (Vertical).* | The amount of detail in the vertical direction of a picture. It is expressed as a number of distinct horizontal alternately black and white lines which can be seen in a picture. |
| *Reverse Current.* | The DC flow resulting when a semiconductor junction is biased in its high resistance or "nonconducting" direction. |
| *R.F. (Radio Frequency).* | Signals with a repetition rate (frequency) above the audible range, but below the frequencies associated with heat and light. Generally taken to be from 20,000 Hz to 30,000,000,000 Hz. (30,000 MHz.) |
| *Ringing.* | An oscillatory transient in a system as a result of a sudden change. |
| *Ripple.* | Amplitude variations in the DC output voltage of a power supply caused by insufficient filtering. |

*Rise Time.*          The time required for a sharp pulse to go from minimum to maximum or vice versa.

*Roll.*          A loss of vertical synchronization which causes the picture to move up or down on a monitor.

*Rolloff.*          Gradual attentuation at either or both ends of the transmission pass band.

*Saturation (Color).*          The vividness of a color, directly related to the amplitude of the chrominance signal.

*Saturation Current.*          The collector current of a transistor flowing with zero emitter current. Sometimes called leakage current or collector cut-off current.

*Scanner.*          A motor powered device for mounting a camera and moving it horizontally. The term is peculiar to the video surveillance field.

*Scanning.*          The rapid movement of the electron beam in the vidicon tube or cathode ray tube in a line-for-line manner across the photo-sensitive surface which produces or reproduces the video picture.

*Scanning.*          When used in the video surveillance field, refers to pan or horizontal camera motion.

*Sequential Switch.*          A switching device that displays the output of several cameras in sequence on a single monitor.

*Selectivity.*          The extent to which a desired frequency can be differentiated from other frequencies.

*Semiconductor.*          A material with electrical resistivities intermediate between those of a conductor (metal) and an insulator. The resistivity may be changed by the application of various physical forces, such as heat, light, or electrical and magnetic fields.

*Setup.*          The difference between the black and blanking level of a television signal.

*Signal-to-Noise-Ratio.*          The ratio between useful TV signal and interference or noise. The higher the

|  |  |
|---|---|
|  | ratio, the better the picture clarity. Noise is a spotty or grainy texture in the picture. |
| *Snow.* | Heavy random noise. |
| *Streaking.* | A condition in which objects appear to be extended horizontally beyond their normal outline. |
| *Sync.* | A contraction of "synchronous" or "synchronize." |
| *Sync Compression.* | Reduction in amplitude of the sync signal with respect to the picture signal. |
| *Sync Level.* | The level of the peaks of the synchronizing signal. |
| *Tearing.* | A picture condition in which horizontal lines are displaced in an irregular manner. |
| *Termination.* | A non-inductive resistor having the same resistance as the characteristic impedance of the cable. |
| *Tilt.* | Movement of a camera in a vertical direction. |
| *Tilt.* | Deviation from ideal low–frequency response when noted as a nonlevel average video signal between fields. |
| *Transducer.* | A device for converting energy from one form to another. |
| *Transient.* | A temporary, non-periodic, current induced in a circuit or device by a change in loading or applied voltages. It may be in the form of a narrow pulse, or a damped oscillatory wave train. |
| *Undershoot.* | Initial transient response to a unidirectional change which precedes the main transition and is opposite in polarity. |
| *Vertical Interval.* | The time of vertical retrace. |
| *Vertical Retrace.* | Return of the electron beam to the top of the picture tube screen or the pickup tube target at the completion of the field scan. |

*Vidicon Burn.*  The retention of an image on the vidicon which shows on the monitor after the camera has been turned away from the image.

*Vidicon Tube.*  The common type of pick-up tube used in CCTV cameras. Works like the retina of the human eye to translate the effect of light striking its photo-sensitive surface into electrical impulses.

*White Clipper.*  A circuit designed to limit white peaks to a predetermined level.

*White Compression.*  Amplitude compression of the signals corresponding to the white regions of the picture.

*White Peak.*  The maximum excursion of the picture signal in the white direction.

*Zener Voltage.*  If a voltage is applied to a semiconductor junction in its high resistance or "non-conducting" direction, very little reverse current will flow until the applied voltage reaches a specific value. At this point, the reverse current will suddenly increase appreciably. Voltage at this point is the maximum reverse voltage that can be applied and is termed the Zener voltage.

*Zoom Lens.*  A lens system that may be effectively used as a wide angle, standard, or tele-photo lens; a variable focal length lens.

# Index